Programming Big Data Applications

Scalable Tools and Frameworks
for Your Needs

Programming Big Data Applications

Scalable Tools and Frameworks for Your Needs

Domenico Talia, Paolo Trunfio,
Fabrizio Marozzo, Loris Belcastro,
Riccardo Cantini & Alessio Orsino

University of Calabria, Italy

World Scientific

NEW JERSEY · LONDON · SINGAPORE · BEIJING · SHANGHAI · HONG KONG · TAIPEI · CHENNAI · TOKYO

Published by

World Scientific Publishing Europe Ltd.

57 Shelton Street, Covent Garden, London WC2H 9HE

Head office: 5 Toh Tuck Link, Singapore 596224

USA office: 27 Warren Street, Suite 401-402, Hackensack, NJ 07601

Library of Congress Cataloging-in-Publication Data

Names: Talia, Domenico, author.

Title: Programming big data applications : scalable tools and frameworks for your needs /
Domenico Talia, University of Calabria, Italy, Paolo Trunfio, University of Calabria, Italy,
Fabrizio Marozzo, University of Calabria, Italy, Loris Belcastro, University of Calabria, Italy,
Riccardo Cantini, University of Calabria, Italy, & Alessio Orsino, University of Calabria, Italy.

Description: New Jersey : World Scientific, [2024] | Includes bibliographical references and index.

Identifiers: LCCN 2023033203 | ISBN 9781800615045 (hardcover) |
ISBN 9781800615052 (ebook) | ISBN 9781800615069 (ebook other)

Subjects: LCSH: Software patterns. | Software frameworks. | Computer programming. | Big data.

Classification: LCC QA76.76.P37 T35 2024 | DDC 005.7/11--dc23/eng/20240130

LC record available at https://lccn.loc.gov/2023033203

British Library Cataloguing-in-Publication Data

A catalogue record for this book is available from the British Library.

For any available supplementary material, please visit
https://www.worldscientific.com/worldscibooks/10.1142/Q0444#t=suppl

Desk Editors: Rosie Williamson/Shi Ying Koe/Logeshwaran Arumugam

Typeset by Stallion Press
Email: enquiries@stallionpress.com

To our students

Preface

This book introduces and discusses models, systems, and programming frameworks developed for processing and analyzing large datasets. In particular, the book provides an in-depth description of the properties and mechanisms of the main programming paradigms for big data analysis, such as MapReduce, workflows, BSP, message passing, and SQL-like models. Moreover, the book chapters describe, through programming examples, the most commonly used frameworks, such as Hadoop, Spark, Storm, and MPI, designed for the analysis of large collections of data.

The web, the Internet of Things, and social media platforms are enabling people to produce and collect huge amounts of digital data originating from many different sources, including blogs, sensors, mobile devices, wearable trackers, satellites, and security cameras. These data, commonly referred to as "big data", are challenging current storage, processing, and analysis systems and capabilities. For this reason, new models, languages, tools, systems, and algorithms are currently being studied, designed, developed, and deployed to effectively collect, store, analyze, and learn from big data.

This book describes and reviews parallel and distributed paradigms, languages, and systems used today to analyze and learn from big data on scalable computers. In particular, the book provides a detailed description of the properties and mechanisms of the main parallel programming paradigms, and through programming examples, it illustrates the most widely used frameworks for big data analysis. Furthermore, the book discusses and compares the different frameworks by highlighting the main features of each of them, their diffusion (the community of developers and users), and the main advantages and disadvantages of using them to implement big data analysis applications. The final goal of this volume is to help designers and developers in programming big data applications by identifying and selecting the best or most appropriate programming tool(s) based on their skills, hardware availability, application domains, and purposes, while also considering the support provided by the developer community. Real programming examples are presented for each programming language/framework to show how big data applications can be designed and implemented.

About the Authors

Domenico Talia is a professor of computer engineering at the University of Calabria, Italy, and an honorary professor at Amity University, India. He is a senior associate editor of *ACM Computing Surveys*, an associate editor of *Computer*, and a member of the editorial board of *Future Generation Computer Systems*, *IEEE Transactions on Parallel and Distributed Systems*, the *International Journal of Web and Grid Services*, the *Journal of Cloud Computing*, *Big Data and Cognitive Computing*, and the *International Journal of Next-Generation Computing*. His research interests include high-performance computing, big data, machine learning, parallel and distributed data analysis, cloud computing, social media analysis, distributed knowledge discovery, peer-to-peer systems, and concurrent programming models. He has authored several books and more than 400 scientific papers.

Paolo Trunfio is an associate professor of computer engineering at the University of Calabria, Italy. In 2007, he was a visiting researcher at the Swedish Institute of Computer Science (SICS) in Stockholm, Sweden. He currently serves as an associate editor of the *Journal of Big Data*, *IEEE Transactions on Cloud Computing*, and *ACM Computing Surveys* and is a member of the editorial boards of several scientific journals, including *Future Generation Computer Systems*, *Big Data and Cognitive Computing*, the *International Journal of Web Information Systems*, and the *International Journal of Parallel, Emergent and Distributed Systems*. His research interests include cloud computing, big data, social media analysis, parallel and distributed knowledge discovery, and peer-to-peer systems.

Fabrizio Marozzo is an assistant professor of computer engineering at the University of Calabria, Italy. He received a PhD in systems and computer engineering at the University of Calabria. In 2011–2012, he visited the Barcelona Supercomputing Center for a research internship with the Grid Computing Research Group in the Computer Sciences Department. He sits on the editorial boards of several journals, including *IEEE Access, IEEE Transactions on Big Data*, the *Journal of Big Data, Big Data and Cognitive Computing, Algorithms, Frontiers in Big Data, Heliyon*, and *SN Computer Science*. His research interests include big data analysis, social media analysis, high-performance computing, cloud and edge computing, and machine learning.

Loris Belcastro is a researcher in computer engineering at the University of Calabria, Italy. He received a PhD in information and communication engineering at the University of Calabria. In 2012, he held a scholarship at the Institute of High-Performance Computing and Networking of the Italian National Research Council (ICAR-CNR). He served as a guest editor for numerous journals, including *Future Generation Computer Systems*, the *Journal of Big Data, Sensors, Algorithms, Applied Sciences*, and *Frontiers in Big Data*. His research interests include cloud and edge computing, big data, social media analysis, and parallel and distributed data analysis.

Riccardo Cantini is a researcher in computer engineering at the University of Calabria, Italy. He received a PhD in information and communication technologies at the same university. During 2021–2022, he was a visiting researcher at the Barcelona Supercomputing Center, working with the Workflows and Distributed Computing Group in the Computer Sciences Department. His research interests include social media and big data analysis, machine and deep learning, natural language processing, opinion mining, topic detection, edge computing, and high-performance data analytics.

Alessio Orsino is currently pursuing a PhD in information and communication technologies at the University of Calabria, Italy. In 2023, he was a visiting researcher at the Department of Computer Science and Technology at the University of Cambridge, collaborating with the Mobile Systems Research Lab. His research interests include big data analysis, parallel and distributed computing, high-performance data analytics, cloud and edge computing, and machine learning.

Acknowledgments

We wish to thank Rosie Williamson and Logesh Arumugam for their useful support and comments on the early drafts of this book and during all the editorial processes. We acknowledge partial financial support from the eFlows4HPC project (European Commission through the Horizon 2020 Research and Innovation program and the EuroHPC JU under contract 955558) and from "PNRR MUR project PE0000013-FAIR" — CUP H23C22000860006.

Contents

List of Figures

List of Tables

Chapter 1

Introduction

Parallel processing architectures and scalable programming languages and tools are playing a key role in the design and implementation of complex software applications that manage and analyze large datasets stored in file systems, databases, archives, and data lakes. This book is written for people who are interested in programming big data applications on multi-core computers, cloud platforms, distributed computing systems, and massively parallel machines. Students, developers, and scholars who are interested in learning about the most effective software frameworks for designing data-intensive applications will find in this book a guide that introduces models, languages, and tools for efficiently managing and analyzing big data and discusses how to select the most appropriate environment to reach the application goals. This chapter introduces the goals of the book, illustrates the topics, and describes its organization.

1.1 Motivation and Goals

The world today generates an unprecedented amount of data, and the ability to extract valuable insights from these data is critical to success in many fields, including business, science, and governance, enabling innovation and informed decision-making. The best way to exploit the value of the massive amount of available data is to implement scalable data management and analysis applications that efficiently extract useful patterns, models, and trends from them. Programming big data applications is a challenging and multifaceted

1

task that requires technical expertise and a deep understanding of various concepts, including data analytics, distributed computing, parallel processing, and machine learning. Despite the significant challenges involved, the big data field is constantly growing, and with it, the demand for skilled professionals who can design and build efficient and scalable applications to handle large amounts of data. The ability to work with big data has become a crucial skill in today's job market, and mastering it can open up numerous career opportunities.

This book aims to be an essential guide for developers looking to develop robust and scalable big data applications. With its comprehensive coverage of the main tools and frameworks, this book offers an in-depth understanding of the principles and practices required to implement efficient big data analysis applications by covering a wide range of topics, including distributed systems, scalable data processing, data management, and machine learning, using popular tools and frameworks, such as Hadoop, Spark, Hive, MPI, and Storm. The book provides a practical approach to building big data applications, making it a suitable guide for developers with different levels of experience.

One of the key benefits of this book is its focus on scalability. In fact, the tools and frameworks discussed here are specifically designed to handle large datasets and run complex processing tasks by exploiting parallelism. Mastering them enables developers to build applications that can handle huge volumes of data. Another significant advantage of the book is its practical approach. The book provides real-world examples that show readers how to apply what they have learned and acquire experience dealing with various practical use cases. Additionally, the book includes a comparison of big data tools for real-world applications that shows how big data are being used in different scenarios and domains, providing readers with a deep understanding of the potential applications of big data and providing guidance on how to choose the right tools for each specific use case. We also cover some of the latest trends, such as exascale computing and parallel and distributed machine learning, and discuss how they can be exploited to analyze and process large datasets.

By the end of this book, the reader will have a thorough understanding of the principles and techniques used to build scalable and reliable big data applications, as well as practical experience with some of the most widely used tools in the field.

1.2 Main Topics

This book provides a comprehensive guide that explores the main paradigms and frameworks used to process and analyze big data, supporting programmers and developers in identifying the best programming tools based on their skills, hardware availability, application domains, and purposes. It covers a broad range of topics related to big data processing, management, and analysis, including:

- The main distributed storage systems, which are essential to face the current exponential growth in data storage requirements, ensuring scalability, efficiency, fault tolerance, availability, and consistency;
- The main principles underlying data analysis and data science processes, as well as their development on scalable computing systems;
- The advantages of technologies such as high-performance computing, cloud computing, and distributed computing, which are helpful in processing large amounts of data in real-world contexts;
- The main programming models for big data, such as MapReduce, workflows, and message passing, which are key paradigms to support users in expressing parallel algorithms and applications by providing an abstraction for a parallel computer architecture;
- The latest proposals in the area of exascale computing, which aim at providing scalable solutions and tools in a wide range of scientific fields, including physics, biology, and simulation of natural phenomena;
- The most used programming tools for big data processing, which offer both general-purpose and tailored solutions to deal with different kinds of data, ranging from structured data to graphs and streams, and domains including batch, streaming, graph-based, and query-based applications;
- The key features, pros, and cons of the different frameworks with respect to specific classes of applications to support programmers in choosing the most appropriate framework, along with other

important factors that can drive this choice, such as data type, infrastructure scale, developer skills, and community size.

1.3   Audience and Organization

This book is intended for students and researchers who are studying big data processing and analytics, software developers, and business professionals who are interested in leveraging big data for their organizations. Readers should typically have a good understanding of programming languages such as Java, Python, or Scala and a basic knowledge of the main concepts of parallel and distributed programming. To cover the needs of such a broad range of readers, the book provides comprehensive overview of programming frameworks for scalable data-intensive distributed and parallel applications. It represents a valuable resource for students seeking a deeper understanding of these concepts and techniques, as well as for professionals working in software factories and data science companies who can benefit from the practical insights and real-world applications presented in the book chapters.

Readers should feel free to tailor their reading experience based on their own skills and familiarity with the topic at hand. They may choose to read the book in its entirety or to focus on specific sections of interest without feeling obligated to carefully read every word before proceeding.

The book has five further chapters, which are outlined below:

Chapter 2: *Big Data Concepts* presents the big data field by introducing the main principles and features of big data management and analysis. In particular, data analysis techniques and data science approaches are discussed, and their development on scalable computing systems is investigated. Technologies such as high-performance computing, cloud computing, and distributed computing are also discussed, explaining how they are useful in big data processing.

Chapter 3: *Programming Models for Big Data* introduces and discusses the main programming models designed and used for implementing large-scale big data applications, including MapReduce, workflows, bulk synchronous parallel (BSP), message passing, SQL-like, and partitioned global address space (PGAS). It also presents the latest proposals in the area of exascale computing. For each programming model, the chapter delves into the key features and mechanisms that can be used to process and analyze big data.

Chapter 4: *Tools for Big Data Applications* describes the programming languages, libraries, and tools used for developing scalable big data applications. Frameworks, including Hadoop, Spark, Storm, and MPI, are presented by describing their programming features and mechanisms. For each programming tool, a few real-world examples of big data applications are put forth.

Chapter 5: *Comparing Programming Tools* compares the programming tools presented in the previous chapter by highlighting the main features, advantages, and disadvantages of using them for different types of applications, such as batch, streaming, graph-based, and query-based applications. It also includes a discussion of the developer community, comparing these frameworks in terms of diffusion and popularity from both end-user and developer perspectives.

Chapter 6: *Choosing the Right Framework to Tame Big Data* concludes the book by discussing the main factors that can influence the choice of the most appropriate framework to process and analyze big data. It mainly focuses on the characteristics of input data, the class of applications, and the infrastructure scale, while also recalling many other factors that can influence, to some extent, the decisions of developers, including the skills of the designer or developer, community size, data privacy, security requirements, available budget, integrability, and interoperability.

1.4 Online Resources

An online repository including all the codes and datasets used in the examples presented in the book chapters is made available to the reader at https://bigdataprogramming.github.io. The repository provides Docker containers for the seamless execution of the proposed examples. A brief guide for installing, compiling, and running the programs is also available.

A copy of slides based on the contents of the book to be used for educational purposes can be found through the publisher's website. Instructions for accessing the slides can be found on page 257.

Chapter 2

Big Data Concepts

This chapter presents the big data field by introducing the main principles and features of big data management and analysis. In particular, data analysis techniques and data science approaches are discussed, and their development on scalable computing systems is investigated. Technologies such as high-performance computing (HPC), cloud computing, and distributed computing are also discussed, explaining how they are useful in big data processing.

2.1 Big Data Principles and Features

In the past few years, the ability to produce and gather data has increased exponentially. In the Internet of Things (IoT) era, huge amounts of digital data are generated by and collected from several sources, such as sensors, mobile devices, web applications, and services. Moreover, with the widespread adoption of mobile devices, millions of people every day use social media and produce huge amounts of digital data that can be effectively exploited to extract valuable information concerning human dynamics and behaviors. Such data, commonly referred to as "big data", contain valuable information about user activities, interests, and behaviors, which makes it intrinsically suited to a very large set of applications.

Nowadays, the term "big data" is often misused, but it is very important in computer science for understanding business and human activities. Several definitions for big data have been proposed in the literature; however, reaching a global consensus about what it means is not easy. Although not explicitly mentioning the term big data, the first definition was proposed by Doug Laney (an analyst at META Group, now Gartner), in a 2001 report (Laney *et al.*, 2001), which suggested that volume, variety, and velocity are the three dimensions of challenges in data management. Subsequently, Gartner proposed a more formal definition (Gartner, Inc., nd): *"Big data is high volume, high velocity, and/or high variety information assets that demand cost-effective, innovative forms of information processing that enable enhanced insight, decision making, and process automation."*

Such a definition makes use of a three-dimensional model to describe big data, also known as the "3Vs" model (i.e., *volume, velocity*, and *variety*). In particular, volume refers to the amount of data generated, velocity to the speed at which such data are generated, and variety to heterogeneity in terms of structure and format of data originating from different sources. Let's discuss these three properties of big data in more detail:

- *Volume* is probably the first property that comes to mind when thinking about big data. Since exabytes of data are created each day, it is now not uncommon for large companies to have even petabytes of data on storage devices. Managing such an amount of data is often a complex challenge, as it requires unconventional storage and management solutions.
- *Velocity* essentially measures the speed at which data arrive. Some data arrive in real time, while others arrive in a delayed manner, in fits and starts, sent in batches. Thus, it could happen that the collection of data originating from different sources and arriving at the same pace could put the collection system in serious difficulty since traditional computing systems don't work on data arriving faster than they can make sense of it. As an example, let's consider a system that collects data from a sensor network composed of thousands of devices that send data at intervals of time in the order of seconds.
- *Variety* refers to the fact that data can be collected from different sources and made available in widely diverse formats, such as video, text, audio, CSV, and PDF. The need to merge or transfer these data into a common format may require great effort and advanced analytical skills to understand these input data and make them manageable and suitable for analysis.

According to Gartner's definition, big data are not only characterized by the large size of datasets but also by their variety (i.e., data from multiple repositories, domains, or types) and the velocity with which they are collected and processed. In fact, we can collect so much digital data from several sources at such a high rate that the volume of data could easily overwhelm our ability to make use of it. This situation is commonly called "data deluge".

The "3Vs" model has been adopted by much of the IT industry and research (e.g., IBM (Zikopoulos *et al.*, 2011) and Microsoft researchers (Meijer, 2011)). However, many other definitions have been proposed to extend the "3Vs" model by introducing other features, such as *value* (Gantz and Reinsel, 2011; Dijcks, 2013), *veracity* (Schroeck *et al.*, 2012), and *complexity* (Agrawal *et al.*, 2012).

In this regard, a 2011 report from the International Data Corporation (IDC) (Gantz and Reinsel, 2011) provided a "4Vs" definition for big data: *"Big data technologies describe a new generation of technologies and architectures, designed to economically extract value from very large volumes of a wide variety of data, by enabling high-velocity capture, discovery, and/or analysis"*. This definition delineates a new feature of big data, i.e., value, which refers to the capability of creating great advantages for organizations, societies, and consumers from data analyses. On the contrary, Beyer and Laney (2012) extended the "3Vs" model by introducing veracity as the fourth "V". Veracity includes questions about the quality, reliability, and uncertainty of captured data and the outcome of the analysis of that data.

Finally, the definition of big data provided by the National Institute of Standards and Technology (NIST) (Chang and Grady, 2015) introduced a new feature of big data, "Variability": *"Big data consists of extensive datasets — primarily in the characteristics of volume, variety, velocity, and/or variability — that require a scalable architecture for efficient storage, manipulation, and analysis"*. Variability is defined as the changes in other data characteristics that produce variance in data meaning in the lexicon, with a potentially huge impact on data homogenization.

From the discussion above, it is clear that finding a common definition of big data is very difficult. In this regard, a brief review of big data definitions has been discussed by De Mauro *et al.* (2015), which aims to define a consensual definition of big data. Despite the multitude of definitions produced over time, the definition of big data has been further extended and adapted to different application contexts. To this end, new models were produced, introducing and combining new "V"s:

- *Virality* refers to the ability of data to convey a message that can reach a large number of people (e.g., by posting it on social media).
- *Visualization* represents the feature of data to be represented graphically, so as to allow the analyst to grasp its meaning at a glance and quickly draw a conclusion.
- *Viscosity* refers to the ability of the information extracted from the data to spark the interest of people and induce them to take an action (call for action).
- *Venue* describes the origin of the data, which can be gathered from multiple, distributed, and highly heterogeneous sources.

2.2 Data Science Concepts

The French mathematician J. Henri Poincarè said that *"Science is built up of facts, as a house is with stones. But a collection of facts is no more a science than a heap of stones is a house"*. Facts must be based on data, and data not only need to be collected but also organized and structured, and relationships among them must be found and built to make them useful in scientific processes. In these operations, computers can play a very important and novel role. The integrated use of computer science methods and technologies and scientific discovery processes have brought science to a new era where scientific methods changed significantly through the use of computational methods, new data management and analysis strategies that created the so-called *e-science* (Bell *et al.*, 2009). A thousand years ago, science was empirical, and its main goal was to describe natural phenomena. A few hundred years ago, its theoretical branch was born, which started to use models and make generalizations. Thanks to computers, in the past few decades, a computational branch of science has been created based on the simulation of complex phenomena. Today, e-science has unified theory, experiments, and simulations, implementing the process of data-intensive scientific discovery. Large sets of data are generated by digital devices and instruments or by simulators. They are then processed by data analysis, and the resulting information/knowledge is stored in computer archives and on the internet. In the e-science framework, specialists analyze data using data science methods. In the book, *The Fourth Paradigm*, Hey *et al.* (2009) argued that this new paradigm does not only represent a shift in the methods of scientific research but also a shift in the way people think. They declared that the only way to deal with the challenges of this new paradigm is to build a new generation of computing systems to manage, analyze, and visualize the data deluge. This new class of computing systems is represented by HPC architectures, cloud computing systems, and large-scale distributed platforms. Jim Gray's last talk is recorded in *The Fourth Paradigm*, and its final paragraph summarizes in a very clear way how computing has changed science: "*I wanted to point out that almost everything about science is changing because of the impact of information technology. Experimental, theoretical, and computational science are all being affected by the data deluge, and a fourth, "data-intensive" science paradigm is emerging. The goal is to have a world in which all of the scientific literature is online, all of the science data are online, and they interoperate with each other. Lots of new tools are needed to make this happen.*"

Actually, a key discipline for this new paradigm that supports scientific discovery through the intensive use of computers, algorithms, and data is data science. Data science combines computer science, applied mathematics, and data analysis techniques to provide insights based on large amounts of data. This discipline improves discoveries by basing decisions on insight extracted from large datasets through the use of algorithms for collecting, cleaning, transforming, and analyzing (big) data. Although data science is a young discipline, its history began in the 1960s. In 1966, Peter Naur coined the term "Datology" as the "*science of nature and use of data*," and in 1974, he used the term data science as the "*The science of dealing with data, once they have been established, while the relation of the data to what they represent is delegated to other fields and sciences.*" However, the first scientific conference to mention data science in its name was the IFCS Conference in Data Science, Classification, and Related Methods held in 1996. The *Journal of Data Science* was established in 2003 to advance and promote data science methods, computing, and applications in all scientific fields.

Data science methods are useful when we have a large volume of data and when patterns are too complex for humans to discover and extract manually. In data science, we may say that algorithms are playing the role of equations. Data gathering and collection are the first steps in data science; however, they are not limited to traditional data collection done in statistics. Data science does not only use traditional data collection methodologies; it also uses data that are already available, i.e., data produced for other goals that are accurately selected and pre-processed and can be used for the analysis of scientific and business processes. In general, having more data helps. However, having the right data is the most important requirement. For this reason, most of the time in the data science processes (around 60–70 percent) is spent collecting and preparing data.

In his survey on data science (Cao, 2017), Longbing Cao summarizes the most commonly used definitions of data scientist. The US National Science Board defines data scientists as *the information and computer scientists, database and software engineers and programmers, disciplinary experts, curators and expert annotators, librarians, archivists, and others, who are crucial to the successful management of a digital data collection.* A report from the US Committee on Science of the National Science and Technology Council defines data scientists as *scientists who come from information or computer science backgrounds but learn a subject area and may become scientific data curators in disciplines and advance the art of data science. Focus on all parts of the data life cycle.* Finally, the Joint Information Systems Committee defined data scientists as *people who work where the research is carried out, or, in the case of data center personnel, in close collaboration with the creators of the data and may be involved in creative inquiry and analysis, enabling others to work with digital data, and developments in database technology.*

2.2.1 *Data science processes*

Data science processes aim at turning a problem into a solution by exploiting data, computational techniques and infrastructures, and analysis techniques. A data science process includes a set of steps as follows:

(1) Framing the problem;
(2) Collecting the needed data;
(3) Processing the data for analysis;

(4) Exploring the data;
(5) Performing in-depth analysis;
(6) Communicating the results of the analysis.

Framing the problem: Understanding and defining the problem is the first step of a data science process. This framing will help data scientists build an effective model that can be appropriate and successful. They need to be able to translate data questions into something actionable. Too ambitious or ambiguous inputs from the people posing the problems must be identified and scarce inputs must be handled to produce actionable outputs. What is important at the end of this stage is having all of the information and context needed to address and solve the problem.

Collecting the needed data: Collecting data is the task of gathering and measuring information on targeted variables of interest for the problem to be solved. Data that allow a data scientist to answer relevant questions and provide future results. The right data are needed to provide the insights needed to turn the problem into a solution. This step of the process involves realizing what data we need and finding ways to gather that data from one or more sources, such as file systems, databases, CRMs, and/or web pages.

Processing the data for analysis: Pre-processing and cleaning raw data is the task of ensuring that the data are in the correct format for the analysis step. Inconsistencies and errors must be identified and dealt with appropriately. Sometimes the data can be messy, containing missing values or errors that may corrupt the analysis. Once the data are cleaned and correctly prepared, they are ready for exploratory data analysis.

Exploring the data: When a large amount of well-structured data are available, exploratory data analysis (EDA) must be run to identify valuable insights that will be useful in the next data science step. Exploratory data analysis is used to look at and analyze datasets and summarize their main features. In this step, data visualization methods may help. It will determine how to process and represent data, making it easier for data scientists to discover patterns, identify anomalies, or test hypotheses.

Performing in-depth analysis: This step is used to create data models, for which data scientists use data mining, machine learning, statistical techniques, and mathematical algorithms to extract insights and predictions from the data. This step is where the data science tools for analysis are used to mine the data and find all possible useful insights in them. Predictive or descriptive data models are produced.

Communicating the results of the analysis: This final step must be implemented to narrate the results of the entire process. Indeed, it is important that users understand why the insights uncovered are important and useful. Proper communication in data science processes will complete and put into action the discovered models and solutions to the addressed problem. In doing so, data scientists may need to exploit visualization tools. Communicating the findings in a clear way will highlight their value.

2.2.2   Data science skills

Data science seems to connect most strongly with areas such as databases and computer science. Several different kinds of skills are needed, including non-mathematical ones. The most relevant are:

- learning the application domain;
- communicating with data owners/users;
- paying attention to data quality;
- knowing how data can be represented;
- managing data transformation and analysis;
- knowing data visualization and presentation;
- considering ethical issues.

Recently, the term "X-informatics" has been coined to represent new research fields where computing techniques and tools are extensively used. One of the most notable is bioinformatics. These fields share common elements: They aim at acquiring new approaches and models by applying informatics-based approaches to existing scientific fields. They also share the data science methodology that includes the generation of large amounts of data and the scalable analysis and discovery of knowledge from those data sources. Other examples of X-informatics are astro-informatics, brain informatics, health informatics, behavior informatics, medical informatics, social informatics, and urban informatics.

Data science requires the management, processing, and analysis of large amounts of data to discover models, trends, and patterns in them. As we will discuss in the book, clouds and HPC infrastructures provide the power that also underpins it by exploiting parallel programming tools and languages. Multidisciplinary skills and knowledge in both HPC and data science help unlock the knowledge contained in the increasingly large, complex, and challenging datasets that are now generated across many areas of science and business. The integration of scalable computing resources, software tools, networking, data, information management, and the machine learning and data analysis algorithms used by data scientists represent the necessary combination for achieving the goal of solving new problems and building new scientific disciplines around data.

The world is progressively moving toward being populated by a data-driven society in which data are the most valuable asset. The proliferation of big data and big computing has boosted the adoption of machine learning and data science across several application domains. Efficient and optimized distributed and parallel in-memory and disk-based execution platforms for complex data analysis jobs are crucial to tackling this challenge. Moreover, data science and data-oriented programming languages are needed for providing complex abstract structures and operations that must be close to problem formulation and data format and organization.

A major issue to be faced in data exploration, processing, and analysis tasks used in data science applications is the time-consuming processes of identifying and training an adequate model. Therefore, data science is a highly iterative, exploratory process in which most scientists work hard to find the best model or algorithm that meets their data challenge. In practice, there are no one-model-fits-all solutions. Indeed, there is no single model or algorithm that can handle all dataset varieties and changes in data that may occur over time. All data analysis and machine learning algorithms require user-defined inputs to achieve a balance between accuracy and generalizability. This task, referred to as parameter tuning, is a long and complex process that must be carefully handled. The exploitation of machine intelligence and data analysis is greatly important for solving complex and challenging problems. In this direction, data science is creating a new era for discoveries.

2.3   Big Data Storage

With the exponential growth of data storage requirements, it became necessary to use distributed storage systems composed of several interconnected servers. In distributed network scenarios, relational databases exhibit scalability limitations that significantly reduce the efficiency of querying and analysis (Abramova *et al.*, 2014). With regard to storage systems, scalability measures the ability of a system to improve performance as the number of storage nodes increases, while ensuring certain other important properties, such as efficiency, fault tolerance, availability, and consistency. Specifically, as depicted in Figure 2.1, two scaling approaches can be adopted (Singh and Reddy, 2015):

- *Vertical scaling*: It consists of increasing the resources (CPU, RAM, disk, and network I/O) of a single server, making it faster and more powerful.
- *Horizontal scaling*: It consists of adding more storage/computing nodes to the system to divide up the workload across them.

In the context of distributed storage systems, most relational databases have little ability to scale horizontally over many servers, which makes it challenging to store and manage the huge amounts of data produced daily by many applications, leading to the need for alternative solutions to store, manage, and query huge amounts

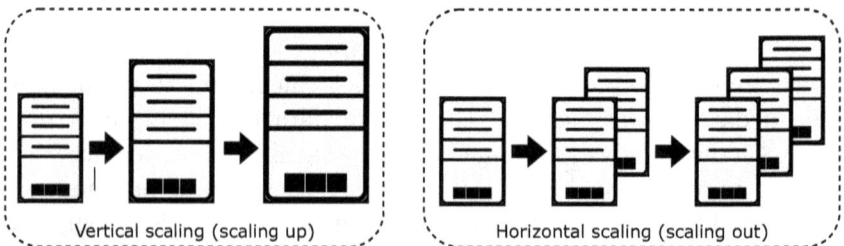

Fig. 2.1.   Vertical vs. horizontal scaling.

of data. In this scenario, "not only SQL" (NoSQL) databases, origi-nally called non-relational databases, became popular in the past few years as an alternative or as a complement to relational databases in order to ensure horizontal scalability of simple read/write database operations distributed over many servers (Cattell, 2011). Compared to relational databases, NoSQL databases are generally more flexi-ble and scalable, as they can transparently take advantage of new nodes without requiring manual distribution of information or addi-tional database management (Stonebraker, 2010). They are also designed to ensure automatic data distribution and fault tolerance (Gajendran, 2012). NoSQL databases provide ways to store scalar values (e.g., numbers, strings), binary objects (e.g., images, videos), or more complex structures (e.g., graphs). According to their data model, NoSQL databases can be grouped into four main storage cat-egories: key–value, document-based, column-based, and graph-based.

Key–value stores provide mechanisms to store data as ⟨key, value⟩ pairs across multiple servers. In such databases, a distributed hash table (DHT) can be used to implement a scalable indexing struc-ture where data retrieval is performed by using a key to find a value.

Document-based databases are designed to manage data stored in documents that use different formats (e.g., JSON), in which each document is assigned a unique key used to identify and retrieve it. Therefore, document stores extend key–value stores as they allow us to store, retrieve, and manage semi-structured information rather than single values. Unlike key–value stores, document stores gen-erally support secondary indexes and multiple types of documents per database and provide mechanisms to query collections based on multiple attribute value constraints (Cattell, 2011).

Column-based data stores (also known as extensible record stores) provide mechanisms to store extensible records that can be parti-tioned across multiple servers. In this type of database, records are said to be extensible because new attributes can be added on a per-record basis. Extensible record stores provide both horizontal parti-tioning (storing records on different nodes) and vertical partitioning (storing parts of a single record on different servers). In some systems, columns of a table can be distributed across multiple servers by using column groups, where predefined groups indicate which columns are best stored together.

Graph-based data stores are widely used systems for storing and querying information that can be represented in the form of graphs instead of tables or documents. In particular, a graph is represented as a set of nodes, edges, and properties, which are difficult to store using a relational database. Graph-based data stores allow to efficiently perform a large set of queries on graphs without the need for costly join operations among tables. In this way, it is possible to speed up the execution of graph algorithms, such as those used for finding communities, degrees, centrality, distances, paths, and other kinds of relationships among nodes.

A brief comparison of NoSQL database management systems is shown in Table 2.1. For a more detailed comparison, see also Hashem *et al.* (2015), Lourenço *et al.* (2015) and Moniruzzaman and Hossain (2013). In the following, the most important NoSQL database management systems are discussed.

2.3.1   *MongoDB*

MongoDB[1] is a modern database management system designed to support internet- and web-based applications. It operates as a document-based NoSQL database, offering a highly efficient data model and persistence strategies that prioritize extensive read and write capabilities. Additionally, MongoDB excels at seamless scalability, enabling effortless expansion with automatic failover mechanisms. One of MongoDB's standout features is its document data model, which simplifies development by providing built-in support for unstructured data. Unlike traditional databases, MongoDB eliminates the need for costly and time-consuming migrations when application requirements evolve.

MongoDB represents documents in a JSON-like format known as BSON, which is a lightweight, fast, and traversable format that fits well with modern object-oriented programming methodologies. BSON serves as the network transfer format for MongoDB's documents. While BSON may initially resemble a binary large object (BLOB), it possesses a crucial distinction: MongoDB comprehends the internal structure of BSON objects. Consequently, MongoDB

---

[1]https://www.mongodb.com.

Table 2.1. Comparison of some NoSQL databases.

	DynamoDB	Cassandra	Hbase	Redis	BigTable	MongoDB	Neo4j
Type	KV	Col	Col	KV	Col	Doc	Graph
Data storage	MEM FS	HDFS CFS	HDFS	MEM FS	GFS	MEM FS	MEM FS
MapReduce	Yes	Yes	Yes	No	Yes	Yes	No
Persistence	Yes	Yes	Yes	Yes	Yes	Yes	Yes
Replication	Yes	Yes	Yes	Yes	Yes	Yes	Yes
Scalability	High	High	High	High	High	High	High
Performance	High	High	High	High	High	High	High, variable
High availability	Yes	Yes	Yes	Yes	Yes	Yes	Yes
Language	Java	Java	Java	Ansi-C	Java Python Go Ruby	C++	Java
License	Proprietary	Apache2	Apache2	BSD	Proprietary	AGPL3	GPL3

Note: FS: file system; MEM: in-memory; KV: key-value; Doc: document; Col: column.

can delve into BSON objects, even when nested, using dot notation. This capability empowers MongoDB to construct indexes and match objects against query expressions, covering both top-level and nested BSON keys.

Beyond its fundamental capabilities, MongoDB boasts comprehensive support for rich queries and full indexes. This sets it apart from other document databases that rely on a separate server layer to handle complex queries. Additionally, MongoDB incorporates convenient features such as automatic sharding, replication, and streamlined storage management. As MongoDB's popularity continues to soar and a wealth of sensitive user information resides within these databases, concerns regarding data confidentiality, privacy, and system security emerge. It is worth noting that when MongoDB was originally designed, security did not receive primary emphasis from its creators.

2.3.2 *Google Bigtable*

Google Bigtable[2] is a popular table store. Built on top of the Google File System (GFS), it is able to store up to petabytes of data and support tables with billions of rows and thousands of columns. Thanks to its high read and write throughput at low latency, Bigtable is an ideal data source for batch MapReduce operations (Chang *et al.*, 2008) and other applications oriented toward the processing and analysis of large volumes of data.

Data in Bigtable are stored in sparse, distributed, persistent, and multi-dimensional tables composed of rows and columns. Each row is indexed by a single row key, and related columns are typically grouped into sets called column families. A generic column is identified by a column family and a *column qualifier*, which is a unique name within the column family. Each value in the table is indexed by a tuple <row key, column key, timestamp>. To improve scalability and balance the query workload, data are ordered by row key, and the row range for a table is dynamically partitioned into contiguous blocks, called *tablets*. These tablets are distributed among different nodes of a Bigtable cluster (i.e., *tablet servers*). To improve load

---

[2]https://cloud.google.com/bigtable/.

balancing, the Bigtable master can split larger tablets and merge smaller ones, redistributing them across nodes as needed. To ensure data durability, Bigtable stores data on GFS and protects it from disaster events through data replication and backup. Bigtable can be used in applications through multiple clients, including *Cloud Bigtable HBase*, a customized version of the standard client for the industry-standard Apache HBase.

2.3.3 *HBase*

With the advent of the internet, huge amounts of structured and semi-structured data began to be generated, encompassing various forms such as emails, JSON, XML, and CSV files. This proliferation of semi-structured data has occurred worldwide, presenting a significant challenge in terms of storage and processing.

Apache HBase[3] has emerged as a rapidly adopted solution to address these challenges. HBase draws inspiration from Google Bigtable, the distributed storage system introduced in Section 2.3.2. Similar to Bigtable's utilization of the distributed data storage provided by the Google File System, Apache HBase leverages Hadoop and Hadoop Distributed File System (HDFS) to provide similar capabilities. Prominent companies, such as Facebook, Netflix, Yahoo!, Adobe, and Twitter, rely on HBase as a core component of their systems. The primary objective of HBase is to accommodate large-scale tables containing billions of rows and millions of columns by utilizing clusters of commodity hardware.

Although HBASE is a column-based store, using a data model that is heavily inspired by Google Bigtable, at its core it operates as a key-value store, where data is stored and retrieved based on unique keys. This key-value model allows for rapid access to specific data points. Each key consists of distinct components, including a row key, column family, column qualifier, and timestamp. This multi-dimensional structure enables the storage of multiple versions of data (using the timestamp), providing a historical perspective on information changes over time. In HBase, row keys are lexicographically sorted, enabling fast range queries, a feature lacking in

---

[3]https://hbase.apache.org/.

relational databases without any sort order assurance. Furthermore, HBase adopts a sparse data storage approach, meaning it doesn't store empty or null values, optimizing storage space.

In HBase, tables are schema-less, and column families are defined during the creation of a table rather than individual columns. The design of an HBase system is intended to scale linearly, consisting of standard tables with rows and columns similar to traditional databases. Every table must have a primary key defined, which must be used for all access attempts to HBase tables. HBase relies on ZooKeeper[4] for high-performance coordination, and it integrates well with Hive, which works as a query engine for batch processing of big data to enable fault-tolerant big data applications. Moreover, HBase can be configured to utilize Apache Avro,[5] a language-agnostic data serialization system that provides a compact binary data format. Avro is designed to be highly efficient for data storage and transmission. By employing Avro, data can be efficiently stored and retrieved in HBase tables, ensuring compatibility across different programming languages and facilitating seamless integration with other systems in a data processing pipeline.

2.3.4 *Redis*

Redis[6] is a popular open-source in-memory data store used as a database, cache, message broker, and streaming engine. It allows you to perform different types of atomic operations, such as appending a string, incrementing the value in a hash, inserting an element in a list, calculating the intersection, union and/or difference of sets, and extracting the element with the maximum score from an ordered set. Although Redis works with in-memory datasets to improve performance, depending on the use case, it can persist data by periodically downloading them to disk.

Redis also includes other interesting features, such as transactions, publish/subscribe services, keys with a limited time-to-live, and automatic failover. It is written in ANSI C and works on most POSIX systems, such as Linux, BSD, and MacOS, without

---

[4]https://zookeeper.apache.org/.
[5]https://avro.apache.org/.
[6]https://redis.io/.

external dependencies. Linux and MacOS are the two operating systems recommended for using Redis, as it is developed and tested mostly on them. Redis can run on Solaris-based systems, though support is limited. Also, there is no official support for Microsoft Windows. However, it is possible to install Redis on Windows for development by using WSL2 (Windows Subsystem for Linux), which enables you to run Linux binaries natively on Windows. Several open-source clients are available for Redis developers, supporting a large set of languages, including Java, Python, PHP, C, C++, C#, JavaScript, Node.js, Ruby, R, and Go.

Since Redis supports many value types and data structures, it is exploited in many use cases, such as broadcasting messages using publish/subscribe technologies, storing random data that require quick access from multiple servers, and managing queues of messages/tasks. Finally, Redis is usually used to store web sessions in order to implement the so-called "sticky sessions", which allow one to maintain a user's login even across different servers hosting the same website. As an example, when accessing Facebook, users may be redirected by the load balancer to a server other than the one on which they are logged in. Using Redis, users will not lose their existing sessions, as the session archive is shared between the different instances of the website.

2.3.5 *DynamoDB*

Amazon DynamoDB is a fully managed NoSQL database service provided by Amazon Web Services (AWS), which is designed to deliver fast and predictable performance with seamless scalability. DynamoDB stores data in a key-value format and is ideal for applications that require low-latency access to data, such as gaming apps, mobile applications, and e-commerce platforms.

One of DynamoDB's key advantages over competitors like Redis lies in its automatic scaling and load balancing capabilities, which eliminate the need for manual intervention in managing database resources. While Redis also offers high performance and low latency, DynamoDB's managed service simplifies operational complexities, allowing developers to focus on building applications rather than managing infrastructure.

DynamoDB provides built-in security features, such as encryption at rest and in transit, fine-grained access control, and integration with AWS Identity and Access Management (IAM). These security measures are crucial for applications dealing with sensitive data, providing a level of confidence to businesses and developers. While other databases offer security features, DynamoDB's integration within the AWS ecosystem enhances its accessibility and ease of management.

DynamoDB offers fully managed backups, restore capabilities, and multi-region, multi-master synchronization, ensuring high availability and data durability. Moreover, it provides high-level integration with other AWS services, such as AWS Lambda, Amazon S3, and Amazon Kinesis, enabling developers to build powerful, serverless architectures. This level of integration streamlines the development process and allows for the creation of highly responsive and cost-effective applications.

In summary, while each NoSQL database has its strengths, DynamoDB's combination of seamless scalability, robust security features, managed services, and tight integration within the AWS ecosystem makes it a compelling choice for developers and businesses relying on AWS infrastructure for their applications.

2.3.6   *Apache Cassandra*

Apache Cassandra[7] is a distributed database management system that provides high availability with no single point of failure. Born at Facebook and inspired by Amazon Dynamo and Google BigTable, Apache Cassandra is designed for managing large amounts of data across multiple data centers and cloud availability zones.

Cassandra uses a masterless ring architecture in which all nodes play an identical role, which allows any authorized user to connect to any node in any data center. This is a really simple and flexible architecture that allows for adding nodes without service downtime. The process of data distribution across nodes is very simple, with no programmatic operations required from the developers. Since all nodes communicate with each other equally, Cassandra has no single point of failure, which ensures continuous data availability and service uptime. Moreover, Cassandra provides a customizable

---

[7]http://cassandra.apache.org/.

data replication service that allows the replication of data across nodes organized in a ring. In this manner, in the event of a node failure, one or more copies of the needed data are available on other nodes. Replication can be configured to work across one data center, many data centers, and multiple cloud availability zones. Focusing on performance and scalability, Cassandra achieves an almost linear speedup, which means the operations per second (OPS) capacity can be increased by adding new nodes (e.g., if two nodes can handle 10,000 OPS, four nodes will support nearly 20,000 OPS, and so on).

Many companies have successfully deployed and benefited from Apache Cassandra, including Apple (75,000 nodes storing over 10 PB of data), Chinese search engine Easou (270 nodes, 300 TB, and over 800 million requests per day), eBay (over 100 nodes and 250 TB), Netflix (2,500 nodes, 420 TB, and over 1 trillion requests per day), Instagram, Spotify, and Rackspace.

2.3.7   *Neo4j graph database*

In many cases, it is required to store data relationships, which refer to the connections or associations between different entities or nodes. A data relationship describes how two nodes are related to each other and can have properties that characterize the relationship. As an example, in a social network graph, nodes might represent users, and relationships might represent connections like 'friend,' 'follow,' or 'married with.'

If we need to take into account real-time data relationships (e.g., create queries using data relationships), NoSQL databases are not the best choice. In fact, relationship-based or graph databases have been created to naturally support operations on data that use data relationships. Graph databases provide a novel and powerful data modeling technique that does not store data in tables but in graph models (Rodriguez and Neubauer, 2010), with several benefits in storing and retrieving data connected by complex relationships.

Among the several graph data models, such as OrientDB, Virtuoso, Allegro, Stardog, and InfiniteGraph, we focus on Neo4j. Neo4j is an open-source NoSQL graph database implemented in Java and Scala and is considered the most popular graph database in use today. The Neo4j source code and issue tracking are available on GitHub, where there is a large support community. Currently, it is used by numerous organizations working in different sectors, including software analytics, scientific research, project management, recommendations, and social networks.

In the Neo4j graph model, each node contains a list of relationship records that refer to other nodes and additional attributes (e.g., timestamp, metadata, and key–value pairs). Each relationship record must have a name, a direction, a start node, and an end node, and can contain additional properties. One or more labels can be assigned to both nodes and relationships. In particular, such labels can be used for representing the roles a node plays in the graph (e.g., user, address, company) or for associating indexes and constraints to groups of nodes.

Moreover, Neo4j clusters are designed for high availability and horizontal read scaling using master–slave replication. Focusing on performance, Neo4j is thousands of times faster than SQL in executing traversal operations. The traversal operation consists of visiting a set of nodes in the graph by moving along relationships (e.g., finding potential friends in a social network based on user friendship). With this operation, graph models allow us to take only the required data into account without doing expensive grouping operations, as done by relational databases during join operations (Vukotic *et al.*, 2015). Queries in Neo4j are written using *Cypher*, a declarative and SQL-based language for describing patterns in graphs. Cypher is a relatively simple but very powerful language that allows us to execute queries easily on complex graph databases.

2.3.8   *Summary considerations about NoSQL storage*

Choosing the best database solution for creating a big data application requires several aspects to be considered. To decide what kind of database to choose, the first aspect to be considered is probably the classes of queries that will be run. So, graph databases are probably the best solution for representing and querying highly connected data (e.g., data gathered from social networks) or those that have complex relationships and/or a dynamic schema. In cases where non-graph data are analyzed, the use of graph databases could result in really poor performance. In this regard, summary considerations on graph databases are presented in Table 2.5.

Another aspect to be considered in choosing the best database solution should be the consistency, availability, and partition tolerance (CAP) capabilities offered, because distributed NoSQL database systems cannot be fully CAP compliant. In fact, the CAP theorem, also named Brewer's theorem(Gilbert and Lynch, 2002), states that a distributed system can't simultaneously guarantee all three of the following properties:

(1) consistency (C), which means all nodes see the same data at the same time;
(2) availability (A), which means every request will receive a response within a reasonable amount of time;
(3) partition tolerance (P), which means the system continues to function even if arbitrary network partitions occur, because communication between nodes is lost or delayed.

Thus, if a distributed database system guarantees consistency and partition tolerance, it can never ensure availability. Similarly, if you need full availability and partition tolerance, you can't have consistency, at least not immediately. In fact, in a distributed environment, data changes on one node need some time to be propagated to the other nodes. During that time, the copies will be mutually inconsistent, which may lead to the possibility of reading not updated data. To try to overcome this limitation, the *eventual consistency* property is usually provided. It ensures that the system, sooner or later, will become consistent. This is a weak property, so if the adopted database system only provides eventual consistency, the developer must be aware that there exists the possibility of reading inconsistent data. As a result, instead of the traditional atomicity, consistency, isolation, and duration (ACID) properties associated with relational databases, the BASE properties have been defined for NoSQL databases (Ganesh Chandra, 2015). While relational databases emphasize ACID attributes to ensure data integrity, NoSQL databases typically follow BASE principles for increased scalability and performance. In particular, BASE principles stand for:

- *Basically available*: It means that the system guarantees availability according to the CAP theorem. In this way, the system always appears to be working, and this is guaranteed by distributing data across several storage systems with a high degree of replication.

- *Soft state*: Because of eventual consistency, the replicas of data may be inconsistent, and the state of the system may vary over time even without any input. By employing a soft state property, the resources loaded during the initial load can be utilized for subsequent requests.
- *Eventually consistent*: When there are no updates in the application for a specified amount of time, the updates propagate.

In contrast with the ACID paradigm, the BASE model puts an emphasis on availability. In fact, since it is difficult to build a database with ACID properties when data are distributed and synchronization is not feasible, consistency and isolation are frequently

Table 2.2.   Summary considerations about key–value databases.

Key–value databases	
Horizontal scaling	Very high scale provided via sharding.
When to use	Simple data schema or extreme speed scenario (e.g., in real time).
CAP tradeoff	Most solutions prefer consistency over availability.
Pros	Simple data model; very high scalability; data can be accessed using query language, such as SQL.
Cons	Some queries could be inefficient or limited due to sharding (e.g., join operations across shards); no API standardization; maintenance is difficult; unsuitable for complex data.

Table 2.3.   Summary considerations about column-based databases.

Column-based databases	
Horizontal scaling	Very high scale capabilities.
When to use	When consistency and high scalability are needed, without using indexed caching front end.
CAP tradeoff	Most solutions prefer consistency over availability.
Pros	Higher throughput and stronger concurrency when it is possible to partition data; multi-attribute queries; data are naturally indexed by columns; support for semi-structured data.
Cons	Greater complexity; unsuitable for interconnected data.

Table 2.4.   Summary considerations about document-based databases.

Document-based databases	
Horizontal scaling	Scale provided via replication and sharding.
When to use	When record structure is relatively small, and it is possible to store all of its related properties in a single document.
CAP tradeoff	In most cases, prefer consistency over availability.
Pros	High scalability and simple data model; generally support for secondary indexes, multiple types of documents per database; nested documents; MapReduce support for ad hoc querying.
Cons	Eventual consistency with limited atomicity and isolation; queries limited to keys and indexes; unsuitable for interconnected data.

Table 2.5.   Summary considerations about graph-based databases.

Graph-based databases	
Horizontal scaling	Poor horizontal scaling.
When to use	For storing and querying entities linked together by relationships; use cases are social networking and recommendation engines.
CAP trade-off	Usually prefer availability over consistency.
Pros	Powerful data modeling and relationship representation; locally indexed connected data; easy and efficient to query.
Cons	Unsuitable for non-graph data.

sacrificed. One of the key concepts behind BASE is that data consistency is a developer concern and should not be handled by the database system. As a conclusion, for each database family, we provide some summary considerations in Tables 2.2–2.5.

2.4   Scalable Data Analysis

The abundance of digitally stored data requires designing efficient methods to analyze them and extract valuable information from them. In particular, there are two main trends, technological and methodological, which must be taken into account:

- *Technological*: There are huge amounts of data collected and warehoused in many repositories distributed all over the world. Data can be collected and stored at high speeds in local databases, from remote sources, or even from our galaxy. Some examples include datasets from the fields of medical imaging, bioinformatics, remote sensing, and several digital sky surveys. As we discussed, this implies a need for reliable data storage, networking, and database-related technologies, standards, and protocols.
- *Methodological*: Huge datasets are hard to understand, and in particular, the data trends and patterns present in them cannot be comprehended by people directly. This is a direct consequence of the increasing complexity of information, mainly its multi-dimensionality. For example, a computational simulation can generate terabytes of data within a few hours, whereas human analysts may take several weeks to analyze these datasets. For this reason, most of the data will never be read by humans and are rather processed and analyzed by computers.

We can summarize the aforesaid as follows: Whereas some decades ago, the main problem was the lack of information, the challenge now seems to be (i) *the very large volume of information* and (ii) *the associated complexity in processing it for extracting significant and useful parts or summaries.*

Nevertheless, the first aspect does not represent a limitation or a problem for the scientific community: Current data storage, architectural solutions, and communication protocols provide a reliable technological base to collect and store such an abundance of data in an efficient and effective way. Moreover, the availability of high-throughput scientific instrumentation and highly inexpensive digital technologies facilitated this trend from both technological and economic viewpoints. On the other hand, the computational power of computers is not growing as fast as the demand for such data computation requires, and this represents a limit to the knowledge that could potentially be extracted. As an additional aspect, we have to consider that storage costs are currently decreasing faster than computing costs, and this trend makes things worse.

To handle this abundance of data availability (whose rate of production often far outstrips our ability to analyze), applications are emerging that explore, query, analyze, visualize, and, in general, process very large-scale datasets: They are named *data analysis applications*. Computational science is evolving toward data analysis applications that include data integration and analysis, information management, and machine learning. In particular, data analysis in large data repositories can find what is interesting in them by using scalable data analysis and data mining techniques. Data analysis in science helps scientists form hypotheses and provides them support for their scientific practices and solving environments, ensuring that the benefits of knowledge that can be extracted from large data sources are obtained. Advanced data analysis techniques and associated tools can help extract information from large, complex datasets that can be useful in making informed decisions in many business and scientific applications, including tax payment collection, market sales, social studies, biosciences, and high-energy physics. Combining big data analytics and machine learning techniques with scalable computing systems will produce new insights in a shorter time.

In the field of data science, the term *data analytics* is often used as a synonym for *data analysis*. But do these two terms have the same meaning? No, there is actually a slight difference between them. According to a popular definition: *"Data Analytics is the science of collecting and examining raw data with the purpose of drawing meaningful conclusions about that information"*. Therefore, although the terms data analysis and data analytics are often used interchangeably, they are slightly different: Data analytics is a broader term that includes data analysis. In fact, data analysis refers to the "process" of preparing and analyzing data for extracting meaningful knowledge. On the other hand, data analytics also includes the tools and techniques used for this purpose, such as data visualization tools or data warehouses.

Since the 1950s, organizations have been using basic data analytics techniques to discover useful information about market trends or insights. In most cases, such kind of analysis was essentially provided to manually examine numbers in spreadsheets with the aim of extracting information that could be used for future decisions. However, today, to give organizations a competitive edge, data analytics must identify insights for immediate decisions. For many years now, most organizations have collected the data — often quite huge amounts — that stream into their businesses for extracting meaningful and valuable information for business (e.g., to make better decisions or plan marketing strategies) and scientific (e.g., to verify or disprove models or theories) purposes.

When data are maintained at geographically distributed sites, the computational power of distributed and parallel systems can be exploited for the analysis of scientific and business data. Parallel and distributed data analysis algorithms are suitable for such a purpose. Moreover, in this scenario, HPC systems and cloud computing platforms provide effective computational infrastructures for implementing scalable data analysis applications and for machine learning from large and distributed datasets (Belcastro *et al.*, 2017). In particular, cloud computing is receiving increasing attention from the industry and research communities, which view this new computing infrastructure as a key technology for solving complex problems and implementing distributed scalable applications (Belcastro *et al.*, 2021b).

Big data analytics refers to advanced data analytics techniques applied to big datasets. These techniques include data mining, statistics, data visualization, artificial intelligence, machine learning, and natural language processing. The usage of big data analytics produces several benefits, especially in large companies, in terms of cost reduction for storing and querying large amounts of data, the effectiveness of decision-making, and the ability to provide services that better meet customers' needs. Some big data analytics application fields are discussed in the following:

- *Text analytics*: It is the process of deriving information from text sources (e.g., documents, social networks, blogs, and websites), which can be used for sentiment analysis, content classification, text summarization, etc.
- *Predictive analytics*: It is the process of predicting future events or behaviors by exploiting models developed using a variety of statistical, machine learning, data mining, and other analytical techniques (Nyce and Cpcu, 2007).
- *Graph analytics (or network analysis)*: It analyzes the behavior of various connected components through the use of network and graph theories, which is useful for investigating structures in social

networks (Otte and Rousseau, 2002). Examples of graph analytics are path, connectivity, community, and centrality analysis.

- *Prescriptive analytics*: It is a form of advanced analytics that examines data or content to optimize decisions about what actions to take in order to maximize profit, given a set of constraints and key objectives. It is characterized by techniques such as graph analysis, simulation, complex event processing, neural networks, recommendation engines, heuristics, and machine learning.

Despite the great benefits discussed above, dealing with big data is not a bed of roses. In fact, the process of knowledge discovery from big data is generally complex, mainly due to the data characteristics previously discussed, which require addressing several issues. Moreover, big data analytics is a continuously growing field where novel and efficient solutions (i.e., in terms of platforms, programming tools, frameworks, and data mining algorithms) spring up daily to cope with the growing scope of interest in big data. In this context, cloud computing is a valid and cost-effective solution for supporting big data storage and executing sophisticated data analytics applications.

Today, many organizations, companies, and scientific centers produce and manage large amounts of complex data and information. Climate, astronomic, and genomic data, as well as company transaction data, are just some current examples of the massive amounts of digital data that must be stored and analyzed to find useful knowledge in them. Such data and information patrimony can be effectively exploited if it is used as a source to produce the knowledge necessary to support decision-making. This process is computationally intensive, collaborative, and distributed in nature. The development of data analysis algorithms and applications for HPC systems and clouds offers tools and environments to support parallel execution strategies for analysis, inference, and discovery from distributed data available in many scientific and business areas. The creation of frameworks on top of HPC and cloud systems is an enabling condition for developing high-performance data analysis tasks and machine learning techniques, and it meets the challenges posed by the increasing demand for power and abstractness arising from complex data analysis scenarios in science, engineering, and business. Indeed, the use of general-purpose data analysis and machine learning techniques and tools may effectively support the analysis of massive and distributed datasets in science, engineering, industry, and business sectors.

2.5   Parallel Computing

Due to their characteristics, big data are intrinsically suitable for a wide range of applications in several domains, ranging from high-energy physics, bioinformatics, and genomics to socio-political or financial modeling and simulations. However, despite their information-rich nature, big data present a number of issues related to their effective and efficient management and processing due to their enormous volume, high heterogeneity, and speed with which they are generated. Indeed, sequential data analysis algorithms are not feasible for extracting useful models and patterns from big data in a reasonable time. For this reason, high-performance computers, such as many- and multi-core systems, clouds, and multi-clusters, along with parallel and distributed algorithms and systems, are required by data scientists to tackle big data issues (Talia *et al.*, 2015).

2.5.1   *Basic concepts and definitions*

The first fundamental distinction useful to understand the different concepts of parallel systems is the difference between two widely used terms, concurrency and parallelism:

- *concurrency*: two or more tasks can be in progress simultaneously;
- *parallelism*: two or more tasks are being executed simultaneously.

It is worth noting that being in progress does not necessarily involve being in execution, so parallelism requires concurrency but not vice versa. In fact, concurrency can also be achieved with no parallelism, as happens with multitasking on a single-core machine. However, in this case, the execution time is not reduced with respect to a sequential execution. Starting with this, *parallel computing* can be defined as the practice of dealing with a problem of size n by breaking it down into $k \geq 2$ parts, which are solved by leveraging p physical processors simultaneously (Navarro *et al.*, 2014). This problem-solving paradigm, also called *divide et impera*, is only applicable if the problem is parallelizable, i.e., it can be expressed as a decomposition into k

distinct sub-problems. As an example, let's consider the scalar product between two vectors $a = [a_1, a_2, \ldots, a_n]$ and $b = [b_1, b_2, \ldots, b_n]$, defined as:

$$a \cdot b = a_1 b_1 + a_2 b_2 + \cdots + a_n b_n = \sum_{i=1}^{n} a_i b_i$$

The computation can be easily parallelized by decomposing the problem into k partial sums (the *divide* step) to be computed on distinct processors simultaneously (the *impera* step). The final result is then obtained by combining the partial sums (which entails the need for *synchronization*). Formally,

$$\sum_{i=1}^{n} a_i b_i = \sum_{i=1}^{j_1} a_i b_i + \sum_{i=j_1+1}^{j_2} a_i b_i + \cdots + \sum_{i=j_{k-1}+1}^{n} a_i b_i,$$

$$1 < j_1 < j_2 < \cdots < j_{k-1} < n$$

In order to correctly design a parallel algorithm, it is crucial to identify the parallel nature of the problem to be solved. Formally, a problem $\mathcal{P}_D$ with domain D can be of two different types:

- *Data-parallel*: D is a set of data elements, and the solution to the problem can be expressed as the application of a function f to each subset of D: $f(D) = f(d_1) + f(d_2) + \cdots + f(d_k)$.
- *Task-parallel*: F is a set of functions, and the solution to the problem can be expressed as the application of the functions in F to D: $F(D) = f_1(D) + f_2(D) + \cdots + f_k(D)$.

2.5.2 *Parallel architectures*

Parallel and concurrent processing systems can take many distinct forms. *Flynn's taxonomy* classifies the different models of parallel systems depending on the multiplicity of the streams of instructions and data that they can manage. There are four combinations that describe the most common parallel architectures, as depicted in Figure 2.2 and discussed in the following (Flynn and Rudd, 1996; Spezzano and Talia, 1999).

Single instruction stream, single data stream (SISD): This class includes the traditional Von Neumann architecture, on which all traditional computers are based, from personal computers to large

INSTRUCTION STREAMS

		one	many
DATA STREAMS	one	**SISD** *traditional Von Neumann single CPU computer*	**MISD** *pipelined computers*
	many	**SIMD** *vector processors, fine-grained data parallel computers*	**MIMD** *multi-computers, multi-processors*

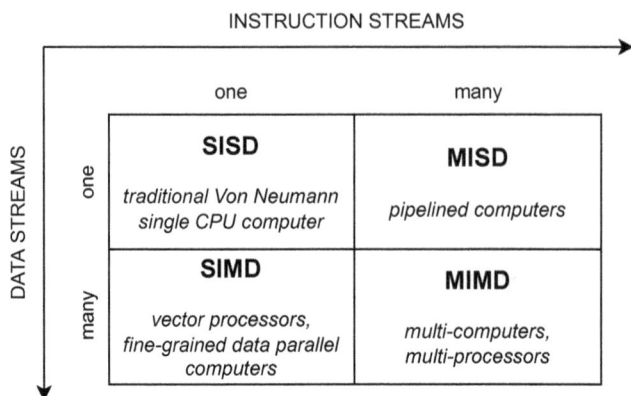

Fig. 2.2.   Flynn's taxonomy.

sequential computers. In such a model, there is a single processor that handles a single flow of instructions executed on a single data flow.

Single instruction stream, multiple data stream (SIMD): This class includes architectures made up of many processing units that simultaneously execute the same instruction but work on different data. Generally, the way to implement SIMD architectures is to have a main processor that sends the instructions to be executed simultaneously to a set of processing elements that will execute them. SIMD systems are mainly used to support the parallel execution of specialized tasks, such as image and signal processing.

Multiple instruction stream, single data stream (MISD): This class is characterized by multiple instruction streams working simultaneously on a single data stream. Abstractly, MISD is a pipeline of multiple independently executing units operating on a single stream of data, forwarding results from one unit to the next. MISD processors have attracted little interest so far, given the absence of established programming constructs that allow programs to easily map onto the MISD model.

Multiple instruction stream, multiple data stream (MIMD): This class, in which several processes are running simultaneously on several processors working on their own or shared data, expresses the most general model and represents an evolution of SISD. Specifically, the realization of MIMD architectures takes place through the interconnection of a large number of conventional processors.

2.5.3 *Hardware platforms*

The scalability of a parallel computing system is a measure of its capacity to reduce program execution time in proportion to the number of its processing elements. According to this definition, scalable computing refers to the ability of a hardware/software parallel system to exploit increasing computing resources effectively in the execution of a software application (Grama *et al.*, 2003). Different platforms implement scaling in various ways to assist large data processing. Specifically, as already discussed (see Figure 2.1), *horizontal scaling* consists of splitting the workload among a large number of independent units (e.g., servers, commodity computers), which are combined to boost processing capability. On the other hand, *vertical scaling* refers to the optimization of a single server, which is made faster by increasing the number of processors (Singh and Reddy, 2015).

Among the main platforms designed to enable horizontal scaling, we can find *client–server architectures* and *peer-to-peer networks*, in which a large number of connected machines (peers) work together in a decentralized and distributed fashion, serving and consuming resources. They are distributed computing platforms in which data exchange and communication between clients and servers or among peers can be implemented by message-passing frameworks, such as sockets or the message passing interface (MPI). For what concerns vertical scaling, instead, we mention *multi-core processors*, *graphics processing units* (GPU), *high-performance clusters*, and *field programmable gate arrays* (FPGA), which are briefly described in the following.

Multi-core CPUs: They are characterized by a high number of processing cores and often use shared memory and a single disk (Bekkerman *et al.*, 2011). The number of cores per chip, as well as the number of operations that a core can do, has recently expanded dramatically, boosting parallelism and enabling the efficient execution of big data analytics algorithms. A specific case of such systems is those constituted by *many-core processors*, which are specially designed to support a high degree of parallelism by leveraging a large number of cores. Compared to multi-core processors, many-core processors have a much higher number of cores (thousands or more) optimized for a higher degree of explicit parallelism and throughput and are therefore widely used in high-performance processing. The main drawbacks of these systems are linked to the physical limitations of the number of processing cores and their dependence on the system memory for data access. Indeed, even if the data fit into the system memory, CPUs can process data at a much faster rate than the memory access speed, which makes memory access a bottleneck.

Graphics processing units: GPUs are optimized hardware components specially designed to accelerate graphical operations (Owens *et al.*, 2008). However, even if GPUs were primarily designed to optimize graphics-related processing, they can also be leveraged for accelerating general-purpose computing. In fact, due to their massively parallel architecture comprising thousands of cores, general-purpose GPUs (GPGPUs) can introduce a significant improvement compared to multi-core CPUs, enhancing extremely parallel applications in many cutting-edge fields, such as deep learning. As an example, the CUDA framework released by Nvidia provides an easy-to-use interface for GPU programming, while many deep learning libraries, such as TensorFlow, allow the programmer to train huge models by fully exploiting GPU power in a completely transparent way.

High-performance clusters: Also called supercomputers, they are machines equipped with a huge number of cores (Buyya, 1999). They rely on powerful hardware which is optimized for speed and throughput and on message-passing communication mechanisms. They are also characterized by a very low error/failure rate because of top-quality hardware; however, this leads to a high monetary cost for deployment and maintenance.

Field programmable gate arrays: FPGAs are highly specialized hardware devices that are designed specifically for certain applications, including medical, video, and image processing, telecommunications, and network security (Brown *et al.*, 1992). FPGAs can effectively be tuned to achieve high speeds, making them orders of magnitude quicker than competing systems. However, the development cost is often substantially greater than other platforms due to custom-built, complex hardware. Also, on the software side, coding is quite complex, as it must be done in *hardware descriptive language* (HDL), which entails a comprehensive understanding of the hardware and, in turn, a considerable cost of the algorithm development process.

2.5.4 *Performance metrics*

There are two main metrics used in parallel computing to measure the performance improvement brought about by the execution of a parallel application, which is generally referred to as scalability. Given T_s, the sequential execution time (on one processor), and T_n, the parallel execution time (on n processors), these metrics are defined as follows:

- *Speedup*: It is the ratio between the sequential execution time of a program and the execution time of the parallel version of that program, defined as $S_n = T_s/T_n$.
- *Efficiency*: It measures the effective use of each processor of a parallel computer when we execute a program, defined as $E_n = S_n/n = T_s/(n \cdot T_n)$.

Amdahl's law can be used to predict the theoretical speedup using multiple processors. It states that the theoretical speedup is limited by the fraction of the program that is actually parallelizable. The remaining portion of the program code can only be run sequentially; therefore, this portion cannot be scaled. Formally, given n the number of processors and F the parallelizable fraction, with $0 \leq F \leq 1$, the theoretical speedup $\hat{S}_n$ is defined as follows:

$$\hat{S}_n = \frac{1}{(1 - F) + \frac{F}{n}}$$

There are several important considerations that can be derived from Amdahl's law, which must be taken into account while designing extreme parallel systems. First, we consider the limit cases for the value of F:

- $F = 0$: This is the worst case, in which no part of a program can be parallelized. This leads to the minimum possible value for $\hat{S}_n$, equal to 1.
- $F = 1$: This is the best case, in which the whole program can be parallelized in every instruction, which leads to the best possible speedup, $\hat{S}_n = n$, which is linear in the number of processors.

Moreover, the maximum speedup obtainable for a program whose parallelizable fraction is equal to F can be derived as follows:

$$S_{\mathrm{max}} = \lim_{n \to \infty} \hat{S}_n = \frac{1}{(1 - F)}$$

From the above formula, it can be easily seen that even if we increase the level of parallelism, the theoretical speedup is bounded by the inverse of $1 - F$, i.e., the non-parallelizable fraction of the program. This has two main implications:

- If the program we are trying to optimize is hardly parallelizable, i.e., $F \approx 0$, the introduction of parallelism will result in a negligible speedup, regardless of the number of available processors.
- Except for the aforementioned special cases ($F = 0$ and $F = 1$), the improvement achieved by raising the level of parallelism follows a diminishing return trend. Specifically, the contribution to $\hat{S}_n$ introduced by the increment of n becomes smaller and smaller as speedup converges to $\frac{1}{(1-F)}$.

All these considerations can be deduced from Figure 2.3.

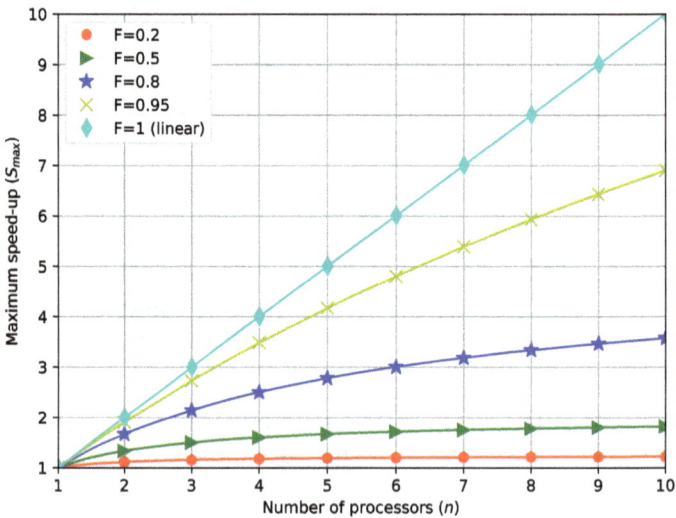

Fig. 2.3.   Maximum speedup obtainable with a varying number of available processors (n), given the parallelizable fraction of the program (F).

2.6 Cloud Computing

In recent years, cloud computing systems have emerged as popular computing platforms to provide effective and efficient data analysis environments to both researchers and companies by tackling the challenge of knowledge extraction from big data repositories in a limited amount of time. From a client perspective, a cloud platform is an abstraction for remote, infinitely scalable provisioning of computation and storage resources. From an implementation point of view, cloud systems are based on large sets of computing resources located somewhere "in the cloud", which are allocated to applications on demand (Barga *et al.*, 2011). Thus, cloud computing can be defined as a distributed computing paradigm in which all the resources, which are dynamically scalable and often virtualized, are provided as services over the internet. As defined by NIST (Mell *et al.*, 2011) cloud computing can be described as: "*A model for enabling convenient, on-demand network access to a shared pool of configurable computing resources (e.g., networks, servers, storage, applications, and services) that can be rapidly provisioned and released with minimal management effort or service provider interaction*".

From the NIST definition, we can identify five essential characteristics of cloud computing systems: *on-demand self-service*, *broad network access*, *resource pooling*, *rapid elasticity*, and *measured service*.

2.6.1 *Cloud service distribution and deployment models*

Cloud systems can be classified on the basis of their service model and their deployment model. Cloud computing vendors provide their services according to three main service models:

- *Software as a Service (SaaS)*, in which software and data are provided through the internet to customers as ready-to-use services. Specifically, software and associated data are hosted by providers, and customers access them without the need to use any additional hardware or software. Examples of SaaS are Gmail, Facebook, Twitter, and Microsoft Office 365.

- *Platform as a Service (PaaS)* is an environment that includes databases, application servers, and development tools for building, testing, and running custom applications. Developers can just focus on deploying applications since cloud providers are in charge of the maintenance and optimization of the environment and underlying infrastructure. Examples of PaaS are Microsoft Azure, Force.com, and Google App Engine.
- *Infrastructure as a Service (IaaS)* is an outsourcing model under which customers rent resources, such as CPUs, disks, or more complex resources like virtualized servers or operating systems, to support their operations (e.g., Amazon EC2, RackSpace Cloud). Compared to the PaaS approach, the IaaS model has higher system administration costs for the user; on the other hand, IaaS presents a higher degree of flexibility, allowing full customization of the execution environment.

The most common models for providing big data analytics solutions in the cloud are PaaS and SaaS. IaaS is usually not used for high-level data analytics applications but mainly to handle the storage and computing needs of data analysis processes. In fact, IaaS is the more expensive delivery model, as it requires a greater investment of IT resources. On the contrary, PaaS is widely used for big data analytics, as it provides data analysts with tools, programming suites, environments, and ready-to-use libraries deployed and run on the cloud platform. With the PaaS model, users do not need to care about configuring and scaling the infrastructure (e.g., a distributed and scalable Hadoop system) because the cloud vendor will do that for them. Finally, the SaaS model is used to offer complete big data analytics applications to end users, so that they can execute analysis on large and/or complex datasets by exploiting cloud scalability in storing and processing data.

Regarding deployment models, cloud computing services are delivered in three main forms:

- *Public cloud*: It provides services to the general public through the internet, and users have little or no control over the underlying technology infrastructure. Vendors manage their proprietary data centers delivering services built on top of them.

- *Private cloud*: It provides services deployed over a company intranet or in a private data center. Often, small- and medium-sized IT companies prefer this deployment model, as it offers advanced security and data control solutions that are not available in the public cloud model.
- *Hybrid cloud*: It is the composition of two or more (private or public) clouds that remain different entities but are linked together.

As outlined by Li *et al.* (2010), users access cloud computing services using different client devices and interact with cloud-based services using a web browser or desktop/mobile app. The business software and user's data are executed and stored on servers hosted in cloud data centers that provide storage and computing resources. Resources include thousands of servers and storage devices connected to each other through an intra-cloud network. The transfer of data between the data center and users takes place over a wide-area network. Several technologies and standards are used by the different components of the architecture. For example, users can interact with cloud services through SOAP-based or RESTful web services (Richardson and Ruby, 2008), while the Open Cloud Computing Interface (OCCI) specifies how cloud providers can deliver their compute, data, and network resources through a standardized interface.

2.6.2 *Cloud services for big data*

At the beginning of the big data phenomenon, only big IT companies, such as Google, Amazon, Yahoo!, Facebook, and Microsoft, invested large amounts of resources in the development of proprietary or open-source projects to cope with big data analysis problems. Today, big data analysis has become highly significant and useful for small- and medium-sized businesses. To address this increasing demand, a large vendor community started offering highly distributed platforms for big data analysis. Among open-source projects, Apache Hadoop is one of the leading open-source data-processing platforms, which was contributed by IT giants such as Facebook and Yahoo!

Since 2008, several companies, such as Cloudera, MapR, and Hortonworks, have started offering enterprise platforms for Hadoop, with great efforts to improve Hadoop's performances in terms of high-scalable storage and data processing. Instead, IBM and Pivotal started offering their own customized Hadoop distributions. Other big companies decided to provide only additional software and support for Hadoop platforms developed by external providers: for example, Microsoft decided to base its offer on the Hortonworks platform, while Oracle decided to resell the Cloudera platform. However, Hadoop is not the only solution for big data analytics. Out of the Hadoop box, other solutions are emerging. In particular, in-memory analysis has become such a widespread trend that companies have started offering tools and services for faster in-memory analysis, such as SAP, which is considered the leading company with its High-performance ANalytic Appliance (HANA)[8] platform. Other vendors, including HP, Teradata, and Actian, have developed analytical database tools with in-memory analysis capabilities. Moreover, some vendors, such as Microsoft, IBM, Oracle, and SAP, stand out from their peers by offering a complete solution for data analysis, including DBMS systems, software for data integration, stream processing, business intelligence, in-memory processing, and the Hadoop platform. In addition, many vendors, such as Amazon Web Services (AWS) and 1010data, decided to focus their entire offering on the cloud. In particular, AWS provides a wide range of services and products on the cloud for big data analysis, including scalable database systems and solutions for decision support. Other smaller vendors, including Actian, InfiniDB, Infobright, and Kognitio, focus their big data offer on database management systems for analytics only.

The remainder of the section introduces the main services for cloud-based big data analytics and machine learning provided by the following popular cloud platforms: Amazon Web Services, Google Cloud Platform, Microsoft Azure, and OpenStack.

---

[8]https://www.sap.com/products/technology-platform/hana.html.

2.6.2.1 *Amazon Web Services*

AWS[9] is a large set of cloud services through which Amazon offers the compute and storage resources of its IT infrastructure to developers. These services can be easily composed by users to build their SaaS applications or integrate traditional software with cloud capabilities. Interaction with these services is facilitated by the presence of SDKs for the main programming languages and platforms (e.g., Java, .Net, PHP, Python, Kotlin, Javascript). The main components and services of AWS are the following:

- *Data transfer*: It includes several data transport solutions designed to securely transfer huge amounts of data into the AWS cloud. These solutions allow for the avoidance of long transfer times, high network costs, and security concerns. They also support the automation of data movement between the AWS cloud and on-premises storage.
- *Data management*: AWS provides a wide variety of database systems, including Amazon Relational Database Service (RDS) for relational tables, and NoSQL solutions, such as Amazon DynamoDB. There are also options for the use of in-memory data caches for real-time applications, provided by Amazon ElastiCache.
- *Compute*: AWS provides Elastic Compute Cloud (EC2) for creating and running virtual servers and Amazon Elastic MapReduce for building and executing MapReduce applications. There are also tools available for automatically scaling system capacity and supporting serverless functions through AWS Lambda.
- *Storage*: AWS provides several flexible storage options for permanent and transient data storage. Among the main storage solutions are the Simple Storage Service (S3), which provides scalable object storage; Amazon Glacier for infrequent access; and Amazon Elastic Block Store, which provides block-level persistent storage.

Among the other components and services that make up the extensive suite provided by AWS, we can find solutions for *networking*, allowing low inter-instance latency and high bandwidth;

---

[9]https://aws.amazon.com.

automation and orchestration, to automate the dynamic scheduling and submission of submitted jobs; and security services, such as AWS Identity and Access Management (IAM). Finally, users can leverage services from the Amazon AI suite to create and deploy applications based on complex artificial intelligence algorithms, including Amazon Polly for text-to-speech translation and Amazon Rekognition for image/face recognition.

2.6.2.2   *Google Cloud Platform*

Google Cloud Platform[10] (GCP) is a wide set of cloud computing services provided by Google, running on the same infrastructure as Google's end-user products (e.g., Gmail). Its growing popularity is due to a number of factors, including efficiency, low latency, and the presence of many innovative tools for big data processing and management, such as BigQuery for data warehousing and Google Cloud Dataflow for real-time data processing. In the following, we list the main services and components provided by GCP.

- *Compute*: This service enables computing and hosting in the cloud. It includes IaaS solutions, such as Google Compute Engine, and several PaaS, including Google App Engine, for developing and hosting web applications in Google-managed data centers and Kubernetes Engine for running containerized applications.
- *Storage*: Storage and database services provided by GCP include Google Cloud Storage and Datastore; SQL-like solutions, such as Cloud SQL; and NoSQL services, such as Bigtable, a wide-column key–value NoSQL database service for large analytical workloads.
- *Networking*: These services enable communication and load balancing across resources. They include Google Cloud DNS, Content Delivery Network (CDN), and security services, such as Armor.

Furthermore, GCP encompasses an extensive set of services in cutting-edge areas, such as big data (with BigQuery and Cloud Data Studio), IoT (with Cloud IoT), and artificial intelligence, with services for supporting a wide range of complex tasks ranging from

---

[10]https://cloud.google.com.

natural language processing and understanding (Cloud Natural Language, Speech-to-Text, Text-to-Speech, and Translation API) to image and video processing (Cloud Vision API and Cloud Video Intelligence).

2.6.2.3 *Microsoft Azure*

Azure[11] is the Microsoft cloud proposal, which is a public PaaS that provides a large set of cloud services, which can be used by developers to create cloud-oriented applications or to enhance existing applications with cloud-based capabilities. The Azure platform provides on-demand compute and storage resources, exploiting the computational and storage power of Microsoft data centers. Azure is designed to support high availability and dynamic scaling services that match user needs with a pay-per-use pricing model. The Azure platform can be used to store large datasets, execute large volumes of batch computations, and develop SaaS applications targeted toward end users. Microsoft Azure includes three basic components/services:

- *Compute*: The computational environment for executing cloud applications. Each application is structured into roles: the web role for web-based applications, the worker role for batch applications, and the virtual machines role for hosting and running virtual machine instances.
- *Storage*: Provides scalable storage to manage binary and text data (Blobs), non-relational tables (Tables), and queues for asynchronous communication between components (Queues). In addition, Microsoft offers its own cloud database services for relational databases, such as Azure SQL Database, which is based on SQL. Furthermore, Microsoft provides Azure Cosmos DB, a NoSQL database service that supports various data models, including key-value, columnar, document, and graph models.
- *Fabric controller*: Aimed at building a network of interconnected nodes from the physical machines of a single data center. The compute and storage services are built on top of this component. Microsoft Azure provides standard interfaces that allow developers to interact with its services. Moreover, developers can use IDEs,

---

[11]https://azure.microsoft.com.

such as Microsoft Visual Studio and Eclipse, to easily design and publish Azure applications.

2.6.2.4 *OpenStack*

OpenStack[12] is an open-source cloud operating system released under the terms of the Apache License 2.0. It is a private IaaS platform that allows the management of large pools of processing, storage, and networking resources in a data center through a web-based interface. Most decisions regarding its development are decided by the community, to the point that every six months a design summit is held to gather requirements and define new specifications for the upcoming release. The modular architecture of OpenStack is composed of four main components, described in the following:

- *Compute*: Provides virtual servers on demand by managing the pool of processing resources available in the data center. It supports different virtualization technologies, such as VMware and KVM, and it is designed to scale horizontally.
- *Storage*: Provides a scalable and redundant storage system. It supports object storage and block storage, which allow storing and retrieving objects and files in the data center.
- *Networking*: This is responsible for managing the networks and IP addresses within the OpenStack environment.
- *Shared services*: Additional services provided to ease the use of the data center, such as identity services for mapping users and services, image services for managing server images, and database services for relational databases.

In several companies and research labs, OpenStack has been effectively used to provide scalable, dynamic infrastructure for big data management and analytics. According to an OpenStack user survey, about 30 percent of users have deployed or tested big data analysis applications. These applications access and analyze multiple data sources, providing valuable insights for most departments within an organization. An OpenStack-based cloud environment offers efficient big data provisioning to support high volumes of analysis and learning requests with quick deployment time.

---

[12]https://www.openstack.org.

2.7 Toward Exascale Computing

In general terms, *high-performance computing* (*HPC*) refers to the use of parallel data processing to deal with large amounts of data and complex calculations. It aggregates computing power from a huge number of nodes, thus granting high speed, scalability, reliability, and energy efficiency. From the application of HPC solutions to data analytics arises the concept of *high-performance data analytics* (*HPDA*), which enables the efficient execution of complex scientific workflows and data-intensive analytics applications.

Machine learning and artificial intelligence applications, including scientific modeling and simulation software systems, need to exploit the power of HPC infrastructures, such as highly parallel clusters, supercomputers, and clouds. However, as parallel computing research and technologies improve, *exascale computing* systems will be used in the next few years to implement scalable big data analysis solutions in science and business (Talia *et al.*, 2022). Exascale is, therefore, the new frontier of HPC and refers to computing systems capable of at least one exaFLOP, which means that they are able to perform at least 10^{18} floating point operations per second (FLOPS). Specifically, an exascale system can be described by different attributes, as detailed in the following (Bergman *et al.*, 2008):

- *Physical attributes*: They are related to the total power consumption and system size (i.e., area and volume).
- *Computational rate*: It is the rate at which a certain type of operation can be executed per second, measured in FLOPS, instructions per second (IPS), and memory accesses per second.
- *Storage capacity*: It measures how much memory is available in various parts of the storage hierarchy, such as main memory, scratch, and persistent storage.
- *Bandwidth rate*: It is the rate at which data relevant to computation can be moved around the system. Bandwidth metrics include local memory bandwidth, checkpoint bandwidth, I/O bandwidth, and on-chip bandwidth.

The standard way to measure performance in terms of FLOPS relies on 64 bit (double-precision floating point) operations per second and the High Performance LINPACK benchmark (Dongarra *et al.*, 1979), comprising a dense $n \times n$ system of linear equations. This benchmark is used by the *TOP500* supercomputer list, which

Table 2.6.   The top three positions of the TOP500 list in June 2023.

Rank	#1	#2	#3
Rmax/Rpeak (PetaFLOPS)	1,194.00/1,679.81	442.01/537.21	309.10/428.70
Name	Frontier	Fugaku	LUMI
Model	HPE Cray EX235a	Supercomputer Fugaku	HPE Cray EX235a
CPU cores	591,872 (9,248 × 64-core Optimized 3rd Generation EPYC 64C @2.0 GHz)	7,630,848 (158,976 × 48-core Fujitsu A64FX @2.2 GHz)	75,264 (1,176 × 64-core Optimized 3rd Generation EPYC 64C @2.0 GHz)
Accelerator cores	36,992 × 220 AMD Instinct MI250X	—	4,704 × 220 AMD Instinct MI250X
Interconnect	Slingshot-11	Tofu interconnect D	Slingshot-11
Manufacturer	HPE	Fujitsu	HPE
Site, country	Oak Ridge National Laboratory, United States	RIKEN Center for Computational Science, Japan	EuroHPC JU European Union, Finland
Year	2022	2020	2022
Operating system	Linux (HPE Cray OS)	Linux (RHEL)	Linux (HPE Cray OS)

ranks and details the 500 most powerful supercomputers in the world (Strohmaier *et al.*, 2015). Table 2.6 reports the top three supercomputers on the June 2023 list. At the top is the Frontier supercomputer, which has been the world's fastest supercomputer and the first real exascale system, reaching 1.102 exaflops. Systems in the TOP500 list are ordered by *Rmax* and *Rpeak* values. Specifically, Rmax is the highest score (in petaFLOPS) measured using the LINPACK benchmarks suite, while Rpeak is the theoretical peak performance of the system (in petaFLOPS).

2.7.1   *Main challenges of exascale systems*

The adoption of exascale systems is a great opportunity to pave the way for applications in a very wide range of application fields, such as weather forecasting, climate modeling, and personalized medicine (Gagliardi *et al.*, 2019). However, their design and implementation are quite complex and require facing a considerable series of challenges (Bergman *et al.*, 2008). In particular, exascale applications need to avoid or strongly limit synchronization, use less communication and remote memory, and handle software and hardware faults that can occur. In the following, we describe these challenges together with the main proposals to overcome them.

Energy: The total consumption limit for these machines should be of the order of 20 MW of thermal design point (TDP), which leads to the need for drastic changes in the process by which hardware has evolved over time. Specifically, energy consumption per computing unit should be considerably reduced to achieve a good balance in performance, memory capacity, and energy cost per FLOP. From an algorithmic point of view, it will be increasingly important to measure the energy-to-solution ratio in order to contain the induced costs of system operation. Given this new trend, APIs should be developed for hardware energy management via software, so that the user can directly manage the energy budget. As an example, if a user needed to minimize the time to solution, it would be possible to increase the frequency of processors, whilst also being aware of the consequent costs.

Concurrency: To limit energy consumption, manufacturers are creating new processors with a lower clock frequency and a higher number of cores. This implies an increasing need for parallelism to decrease the time-to-solution value. There are different levels of parallelism, such as intra-node and inter-node, i.e., within and between HPC nodes, respectively. Vectorization can also be used at the core level by creating an additional level of intra-core parallelization. Moreover, accelerators can be leveraged to increase single-node performance, achieving extreme intra-core parallelism. Machines that rely on this technology are called hybrid architectures. Future exascale machines must therefore be able to fully exploit the advantages offered by these types of parallelism, minimizing the overhead and the need for synchronization as well.

Data locality: A key aspect of future exascale systems, characterized by heterogeneous and complex architectures, is the minimization of inter-node communication and data movement. Indeed, any computation requested by a data-intensive application is much more efficient if it is executed near the data it operates on. In order to take advantage of data locality, exascale systems should have an appropriate topology. On the other hand, applications should be able to read data locally, being aware of the topology of the nodes on which tasks are executed and the location of the data to be processed. Based on the concept of a rack, i.e., a supporting framework that holds hardware modules, typically servers, hard disk drives, and other computing equipment, several levels of data locality can be defined:

- *Node locality*: This is the best scenario, in which communication is not necessary because tasks run on the same node where the data they operate on are located.
- *Intra-rack locality*: In this scenario, data are stored in a different node of the rack, which implies a need for moving the task to that location. However, this is achievable with low latency, as communication takes place within the same rack.
- *Inter-rack locality*: This is the worst scenario, as tasks must be moved to nodes located on different racks, which can cause performance degradation due to communication overhead.

Summing up, the aim of this strategy is to remove the bottleneck created by data movement. Indeed, besides computational time, an impactful overhead component in exascale systems is due to communication and I/O operations. Consequently, specific data placement policies and data-aware schedulers are needed to enable full exploitation of the underlying hardware and software stacks, following a data-centric approach. In such an approach, data availability among neighbor nodes dictates the operations taken by those nodes, thus limiting the data exchange overhead in massively parallel systems via data-driven local communication (Talia, 2019).

Memory: Due to the aforementioned energy limits, there will be an overall increase in the main memory, but not such as to guarantee an equal increase in memory per single core. Therefore, except for a drastic change in the architecture of the memories, this will lead to a decrease in memory per core since the number of cores must increase considerably to reach the computational power of one exaFLOP. Consequently, algorithms running on such architectures will have to adapt to these constrained resources by minimizing memory usage as well.

Resilience: Given the increase in the number of components in exascale systems, an increase in the number of failures is also expected. Due to this, it is necessary that applications adopt countermeasures, leveraging mechanisms such as checkpoint and restart. The checkpoint mechanism is usually expensive both in terms of time and energy due to the synchronization of computing units and I/O operations. Therefore, in exascale machines, checkpoint time could be even longer than the elapsed time between two failures, which makes the restart unusable. This problem will have to be addressed both from an architectural and algorithmic point of view in order to develop resilient exascale systems.

2.8 Parallel and Distributed Machine Learning

Often, machine learning techniques use centralized execution approaches on one computer or in a data center both for training and model execution. These approaches are not appropriate when the data are very large or are located on many different storage devices. Moreover, when big data sources need to be analyzed, sequential execution times can be very long, taking days or weeks to complete. The most effective approach to reducing execution time is to use parallel and distributed computing models and infrastructures. Parallel computers, such as multi-clusters, clouds, and exascale systems, if equipped with distributed and parallel machine learning tools and algorithms, provide a scalable way to analyze big datasets and obtain results in a reasonable time. This strategy is based on the management of distributed data and the exploitation of the inherent parallelism of most data analysis and machine learning algorithms.

2.8.1 *Parallel learning strategies*

Three main strategies can be identified for the exploitation of parallelism in machine learning algorithms:

- *Independent parallelism* is exploited by running processes in parallel in a fully independent way. Each process accesses the whole dataset or its own partition and does not communicate or synchronize with other processes during the training and learning operations.
- *Single program multiple data (SPMD) parallelism* runs in parallel a set of processes that execute the same algorithm on different partitions of a dataset; SPMD processes cooperate by exchanging partial results during their execution.

- *Task parallelism* is the most general form of parallelism since each process can execute different algorithms on (a different partition of) the dataset; processes may communicate according to the different ways the parallel algorithm requires.

These three basic models, if needed, can be combined to implement hybrid parallel and distributed machine learning algorithms. Here, we describe how the three forms of parallelism can be used to design and run parallel classification, clustering, and associative learning algorithms.

Among classification algorithms, decision trees are a popular and effective technique that exploits tree-shaped structures to classify data items. Paths in those trees, from the root to a leaf, correspond to rules for classifying a dataset, whereas the tree leaves represent the classes and the tree nodes represent attribute values.

Independent parallelism can be exploited in decision tree construction by assigning a process the task of constructing a decision tree according to some parameters. If several processes are executed in parallel on different computing nodes, a set of decision tree classifiers can be obtained at the same time. One or more of such trees can be selected as classifiers for the data.

In the exploitation of task parallelism, one process is associated with each sub-tree of the decision tree that is built to represent a classification model. The search occurs in parallel in each sub-tree; thus, the degree of parallelism P is equal to the number of active processes at a given time. This approach can be implemented using a farm parallelism pattern in which one master process controls the computation and a set of workers that are assigned to the sub-trees. The result is a single decision tree built in a shorter time with respect to sequential tree building.

In SPMD parallelism, a set of processes execute the same code to classify the data items belonging to a subset of the global dataset. The n processes search in parallel in the whole tree using a partition D/n of the dataset D. The global result is obtained by exchanging partial results among the processes. Note that the dataset partitioning may be performed in the following two main ways: (i) by partitioning the tuples of the dataset D by assigning D/n tuples per processor or (ii) by partitioning the k attributes of each tuple and assigning k/n attributes of all tuples to each processing node.

Parallelism in clustering algorithms can be exploited both in the data clustering technique and in the computation of the similarity or distance among the data items by computing on each processor the distance/similarity of a different partition of items. In the parallel implementation of clustering algorithms, the three main parallel strategies described earlier in this section can be applied. In independent parallelism, each processor uses the whole dataset D and implements a different clustering task based on a different number of clusters k_i. To get the load among the nodes balanced until the clustering task is complete, a new clustering task is assigned to a processor that has completed its assigned grouping.

According to task parallelism, each processor executes a different task that executes the clustering algorithm and cooperates with the other nodes by exchanging partial results. For example, in partitioning methods, computing nodes can work on disjoint regions of the search space using the whole dataset. In hierarchical methods, a processor can be responsible for composing one or more clusters. It finds the nearest neighbor cluster by computing the distance between its cluster and the others. Then, all the local shortest distances are exchanged to find the global shortest distance between two clusters that must be merged. The new cluster will be assigned to one of the two processors that handled the merged clusters.

Finally, in SPMD parallelism, each node runs the same algorithm on a different partition D/n of the dataset for computing partial clustering results. Local results obtained on the assigned partitions are then shared among all the processors to compute global values on every processor. Global values are used in all processors to start the next clustering step until convergence is reached or a given number of clustering steps are performed. The SPMD strategy can also be used to implement clustering algorithms, where each node generates a local approximation of a model that, at each iteration, can be sent to the other nodes, which, in turn, use it to improve their clustering model.

Association rule algorithms are often used for discovering complex associations in a dataset. Let D be a set of transactions, the task of mining association rules is to generate all association rules that have support (how often a combination occurred overall) and confidence (how often the association rule holds true in the dataset) greater than the user-specified minimum support and minimum confidence, respectively. Independent parallelism can be exploited to run in parallel association rule algorithms that avoid data dependencies among the different processes. This can be done by partitioning and replicating data and candidate frequent itemsets so that processes can run autonomously. For example, the candidate distribution method (Agrawal and Shafer, 1996) proposed for implementing the Apriori algorithm, partitions candidate itemsets but selectively replicates instead of partitioning and exchanging the database transactions so that each process can proceed independently.

In SPMD parallelism, the dataset D is partitioned among the n nodes, whereas candidate itemsets I are replicated on each processor. Each process p counts in parallel the partial support S_p of the global itemsets on its local partition of the dataset of size D/n. At the end of this phase, the global support S is obtained by collecting all local supports S_p. While the replication of candidate itemsets minimizes communication, it does not use memory efficiently. However, due to low communication overhead, scalability is good.

According to task parallelism, both the dataset D and the candidate itemsets I are partitioned on each processor. Each process p counts the global support S_i of its candidate itemset I_p on the entire dataset D. After scanning its local dataset partition D/n, a process must scan all remote partitions for each iteration. The partitioning of the dataset and the candidate itemsets minimizes the use of memory. However, it requires high communication overhead in distributed memory architectures, as a result of which this approach is less scalable than the previous one.

2.8.2 *Distributed learning strategies*

Distributed data analysis and machine learning techniques use distributed computing systems connected through the internet to store datasets and run algorithms, exploiting their inherent parallelism by accessing data on different computing nodes and running the algorithms locally on those nodes. This approach is suitable for applications that typically deal with very large amounts of data that are located on different servers and cannot be analyzed on a single computer or site using traditional machines within acceptable times. Distributed learning strategies are designed to process data sources located at remote sites, such as web servers or departmental data owned by large enterprises, or data streams coming from sensor networks, social media, or satellites.

Sequential machines are not appropriate for most of the distributed and ubiquitous learning applications that require analyzing big data sources. On the other hand, the long response times, the lack of proper use of distributed resources, and the basic features of centralized data analysis algorithms do not work well in distributed (and parallel) environments. Therefore, as discussed for parallel data analysis techniques, scalable solutions for distributed applications call for distributed processing of data, controlled by expert data scientists taking the available resources into account.

In most distributed algorithms, the same code runs on each site concurrently, and a local model per site is computed. Then, all local models are aggregated/combined at a central site or shared across all nodes to produce the global model. This schema is common to several distributed machine learning algorithms. Among them, ensemble learning, meta-learning, federated learning, and collective data mining are the most common. Moreover, we must mention that distributed algorithms can integrate servers or clusters that, on a single site, run parallel algorithms designed according to the described parallel approaches. This combination can be exploited to implement very large distributed data analysis and machine learning applications where the local models that also compose the global model are computed in parallel according to the techniques discussed in the previous section.

The meta-learning technique aims at implementing a global model analyzing a set of distributed datasets. Meta-learning can be defined as learning from learned knowledge (Prodromidis *et al.*, 2000). In a classification scenario, it is achieved by learning from the predictions of a set of base classifiers on a common validation set. The initial training sets are given as input to N learning algorithms that run on different nodes to build N classification models (base classifiers). A meta-level training set is built by combining the predictions of the base classifiers on a common validation set. Finally, a global classifier is trained from the meta-level training set by a meta-learning algorithm. Stacking is a way of combining multiple models in meta-learning; it is applied to models built by using different learning algorithms. Stacking is used to learn which classifiers are reliable through another learning algorithm (the meta-learner) to discover how best to combine the output of the base learners.

Ensemble learning aims to improve model accuracy by aggregating predictions produced by a set of learners (Tan *et al.*, 2016). An ensemble learning method constructs a collection of classifiers from training data and implements classification by a voting strategy (in the case of classification) or by averaging values (in the case of regression) from the predictions generated by each classifier. The final result is an ensemble classifier that very often shows higher classification accuracy with respect to each base classifier that has been used to compose it. The main stages that compose an ensemble learning strategy for data classification include the use of a partitioning tool that splits the input dataset into a training set and a test set. Then, the training set is given as input to N classifiers, which run on different nodes to build N independent models. Finally, a voter tool V will access the N models and carry out an ensemble classification by assigning to each data instance of the test set the class predicted by the majority of the N models produced at the previous stage. Identifying good or optimal strategies to assemble the base classifiers is a crucial issue. The most commonly adopted approaches are boosting and bagging. Bagging, which is also called voting for classification and averaging for regression, combines the predicted classifications (predictions) from a set of models or from the same type of model for different learning datasets. Bagging is also used to solve the inherent instability of results when complex models are applied to relatively small datasets. Although boosting also merges the decisions of different models, like bagging, by amalgamating the various outputs into a single prediction, it computes the single models in different ways. In fact, in bagging models receive equal weight, whereas in boosting weighting is used to give more influence to the more successful ones.

Differently from other approaches, collective data mining builds the global model through a combination of partial models computed at different sites instead of combining a set of complete models generated at each site on partitioned or replicated datasets. In collective data mining, the global model is directly composed by adding an appropriate set of basis functions. The global classification is based on the fact that any function can be expressed in a distributed fashion using a set of appropriate basis functions that can contain nonlinear terms. If the basis functions are orthonormal, a local analysis generates results that can be effectively used as components of the global model. If a nonlinear term is present in the summation function, the global model is not fully decomposable among local sites, and cross-terms involving features from different nodes must be taken into account. Kargupta *et al.* (1999) discussed the following main steps of a collective data mining strategy:

- An appropriate orthonormal representation is identified for the class of data model to be generated.
- At each node, approximate orthonormal basis coefficients are generated.
- If the global function includes nonlinear terms, a sample of datasets is moved from each node to a central node, and the approximate basis coefficients corresponding to such nonlinear terms are computed there.
- The local models are combined to generate the global model.

The concept of federated learning was first introduced by Google in 2017. It is intended to analyze distributed raw data without being moved to a single server or data center. This strategy selects a set of nodes and sends the first version containing the model parameters of a machine learning model to all the nodes. Then, each node executes the model, trains it only on local data, and maintains a local version of the model. Federated learning enables mobile devices to collaboratively learn a shared learning model while keeping all the training data onboard, thus improving security and privacy. It works in the following ways: Each device downloads a learning model from a server, improves it by learning from local data stored on it, and then summarizes the changes as a small, focused update. This update to the model is sent to a central server or to the cloud, where it is averaged with other local updates to improve the global shared model. All the training data remain on the device, and no individual updates are stored on the server, thus ensuring data privacy. To extend the original federated learning strategy based on a central server, decentralized federated learning approaches have been proposed. In decentralized strategies, the central server is removed, and the nodes are able to coordinate themselves to compute the global learning model. This decentralized strategy avoids single-point failures, as the model updates are shared only among interconnected nodes without the orchestration of a central server. Nevertheless, a high communication overhead may be registered due to a larger number of messages among the nodes. This may affect the performance of the global learning task.

Chapter 3

Programming Models for Big Data

This chapter introduces and discusses the main programming models designed and used for implementing large-scale big data applications. In particular, we describe the main features of popular programming models such as MapReduce, workflows, BSP, SQL-like, and PGAS.

3.1 Parallel Programming for Big Data Applications

Since the real world is intrinsically parallel, parallel calculations are natural. Writing sequential programs often entails imposing an order on independent tasks that may be performed concurrently. Parallel machines, on the other hand, provide greater computational power than a single processor.

3.1.1 *The need for parallel programming models*

A programming model is an interface that separates high-level properties from low-level ones. It provides specific operations for the programming level above and requires implementations for all the operations on the architectural level below (Skillicorn and Talia, 1998a).

A parallel programming model is an abstraction for a parallel computer architecture that aids in the expression of parallel algorithms and applications. It can represent a variety of problems for various parallel and distributed systems. Parallel programming models are often the core feature of big data frameworks since they impact the execution paradigm of big data processing engines as well as the way users design and create applications (Wu *et al.*, 2017). They enable the separation of software development issues from parallel execution concerns while also providing *abstraction* and *stability*. Abstraction is guaranteed because the model's operations are at a higher level than those of the underlying architectures. This simplifies the software structure and the difficulty of its development while guaranteeing stability through a standard interface. Therefore, the model can lower the implementation effort (e.g., compiler and runtime system), as implementation decisions are determined once for each target system rather than making decisions for each program (Skillicorn and Talia, 1998a).

Computation models, which define the rules, principles, and processes by which computations are carried out, are strongly connected to parallel programming paradigms. The primary distinction is that a programming model requires practical concerns about hardware and software implementation, while a parallel computing model does not have to be practical. A variety of big data programming models have been proposed, each with a particular focus and set of advantages.

3.1.2 *Programming model features*

The capacity to express high-level and low-level programming mechanisms distinguishes programming models based on their level of abstraction (Talia, 2019). A programmer uses high-level scalable models to specify only the high-level logic of an application while hiding the low-level details that are not required for application design, such as infrastructure-dependent execution details. With these models, a programmer is supported in the creation of an application, and the performance of the application is dependent on the compiler, which evaluates the code of the application, optimizing its execution. On the other hand, low-level scalable models enable programmers to interface directly with the compute and storage units that comprise the underlying infrastructure and thus specify the application parallelism directly. The following sections focus on programming models as distinct from programming systems (Gropp and Snir, 2013; Belcastro *et al.*, 2022). A programming system, in fact, is an implementation of one or more models, which can be developed in a variety of ways (Talia, 2019), such as:

- With a general-purpose or domain-specific language, or a language extension, which enables the expression of parallelism in applications. This strategy entails the need for devising new parallel programming languages or the integration of a comprehensive set of parallel constructs and data structures into existing languages.

- With annotations in the application code that tell the compiler which instructions must be executed concurrently. Parallel statements are distinguished from sequential constructions in this approach, and they are explicitly identifiable in the program code by specific symbols or keywords.
- By including a library in the application code to enhance parallelism in the application. This is now the most popular approach since it is orthogonal to host languages.

Therefore, for example, MapReduce and message passing are programming models, whereas Apache Hadoop and the MPI are programming systems that support those models.

Given the wide range of big data applications and classes of users, considering both skilled programmers and end users, we examine parallel programming models at various (high and low) levels of abstraction.

3.2 The MapReduce Model

The MapReduce programming model was developed by Google (Dean and Ghemawat, 2004) in 2004 to face the challenge of processing big data effectively. It supports extremely scalable and fault-tolerant distributed computations, including data processing on very large input datasets.

3.2.1 *Key ideas behind MapReduce*

The MapReduce paradigm was inspired by the *map* and *reduce* functions available in functional programming languages, such as LISP, and it allows designers to create distributed applications based on those two operations (Marozzo *et al.*, 2012). Programming interfaces are defined in functional programming as functions that are applied to input data sources, and the computation is handled as a function calculation.

The *divide and conquer* strategy is another essential concept behind the MapReduce model. Specifically, a feasible strategy for dealing with large-data issues is to (i) divide the problem into smaller sub-problems, (ii) execute independent sub-problems in parallel using several workers, and (iii) combine intermediate results from each individual worker. In a general sense, workers can be threads in a processor core, cores in a multi-core processor, multiple processors in a computer, or many machines in a computer cluster.

3.2.2 *The programming model*

The programmer defines two phases for MapReduce processing: map and reduce. The *map* function takes a (key, value) pair as input and produces a list of intermediate (key, value) pairs:

$$map \ (k1, v1) \rightarrow list(k2, v2)$$

The *reduce* function merges all intermediate values with the same intermediate key:

$$reduce \ (k2, list(v2)) \rightarrow list(v3)$$

Parallelism is achieved in both phases: In the map phase, where keys can be processed concurrently by different computers (map calls are distributed across computers by sharding the input data), and in the reduce phase, where reducers working on distinct keys can be executed concurrently. As a consequence, MapReduce algorithms have been demonstrated to scale from a single server to hundreds of thousands of servers. Moreover, the MapReduce approach hides the details of the underlying parallelization from the programmer, making it simple to use. This is one of its main advantages, as only the map and reduce functions are required. Developers concentrate on *what* computations must be performed rather than *how* those computations are carried out or how data are sent to the processors.

As an example of a MapReduce application, we consider the generation of an inverted index (Sarkar *et al.*, 2015). Given a large corpus of documents, this index contains a set of words (or index terms), specifying the IDs of all the documents that contain each word. A MapReduce approach can be effectively leveraged in this case, where the map function generates a sequence of <word, documentID> pairs for each input document. The reduce function takes all the pairs for a given word, sorts the corresponding document IDs, and emits a <word, list (documentID)> pair. The inverted

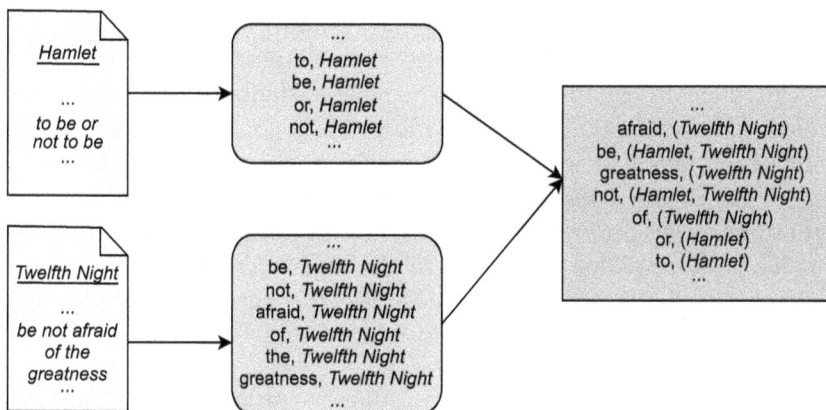

Fig. 3.1.   Inverted index using MapReduce.

index for the input documents is formed by the set of all output pairs created by the reduce function, as illustrated in Figure 3.1. An example application for creating an inverted index for a large set of web documents, based on the MapReduce model, is presented in Section 4.2.1.4.

3.2.3   *MapReduce programs*

3.2.3.1   *Map and reduce phases*

A *job* is a MapReduce program consisting of the code for the map and reduce phases, setup settings (e.g., where output data must be stored), and the input dataset, which is stored on the underlying distributed file system. In fact, for traditional computing systems such as data centers or computer clusters, distributed file systems are the most popular solution for accessing input/output data in MapReduce systems. This is because MapReduce can analyze very large datasets, ranging from gigabytes to petabytes; thus, the input will typically not fit on a single computer's memory.

Each MapReduce job is divided into smaller units known as *tasks*. Map tasks are called *mappers*, while reduce tasks are called *reducers*. To define complex applications that cannot be written with a single MapReduce job, users may need to compose workflows of MapReduce jobs. These workflows can be easily implemented since the output of a job is typically written to a distributed file system and can be used as input for the next job. As a result, a MapReduce program may involve multiple rounds of map and reduce operations.

Current MapReduce systems, such as Hadoop, are built on a master–worker model. A user node sends a job to a master node, which identifies idle workers and assigns each one a map or reduce task. The master coordinates the entire MapReduce job flow, managing both map and reduce tasks. Once all tasks are completed, the master node provides the result to the user node. In detail, the following steps may be taken to describe the entire processing in a MapReduce application (see Figure 3.2):

(1) A job descriptor is sent to a master process, describing the MapReduce task to be done and other information, such as the input data location.
(2) The master starts several mapper and reducer processes on different machines based on the descriptor. It also gathers input data from its location and divides it into several partitions, which are distributed to the different mappers.
(3) Each mapper process uses the *map* function (given as part of the job descriptor) to build a list of intermediate (key, value) pairs after receiving its data partition (i.e., a chunk of data). The code specified by the user for the map function determines how (key, value) pairs are generated from the input data.
(4) The same reducer process is allocated to all pairs with the same keys. As a result, each reducer process runs the *reduce* function

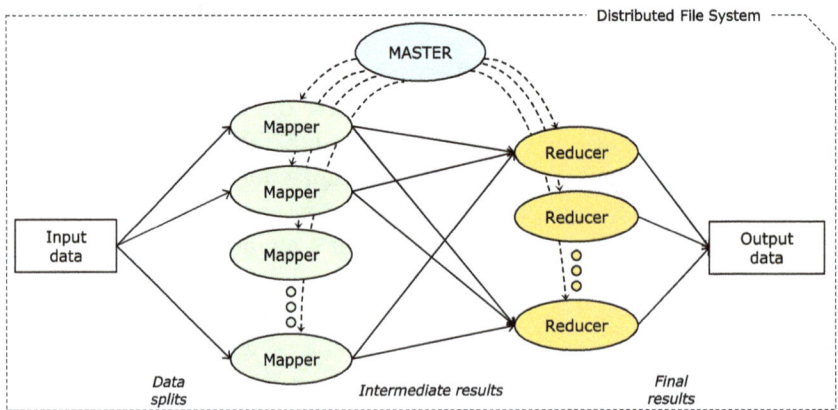

Fig. 3.2.   MapReduce execution flow.

(specified by the job descriptor), which merges all the data associated with the same key to produce a potentially smaller set of values. The user's code for the reduce function determines the manner in which values are combined.

(5) The outputs of each reducer process (r files, where r is the number of reducers) are then gathered and sent to the location specified by the job descriptor, forming the final output data.

3.2.3.2 *Combine phase*

In the execution flow of a MapReduce program, reduce tasks cannot start until the whole map phase is completed, posing a barrier to performance. To increase speed, a combine step can be performed, involving a mini-reduce phase on the local map output. This step performs partial aggregation of data before transmitting them to the reducers over the network. This can occur only if the reduce function is associative and commutative: Values can be combined in any order and yet provide the same result. In this scenario, we use a *combiner* to aggregate local map output:

$$combine \ (\text{k2}, \text{list(v2)}) \rightarrow \text{list(v3)}$$

In many cases, the same function can be used for both combining and the final reduction, with the benefits of reducing the amount of intermediate data and the network traffic.

3.2.3.3 *Shuffle and sort phase*

Between the map (with combine) and reduce phases, an implicit distributed *group by* operation occurs on intermediate keys. This process, known as *shuffle and sort*, passes mapper output to reducers, merging and sorting the output so that intermediate data are sorted by key and arrive at each reducer. Because intermediate keys are not stored on the distributed file system, they are spilled to the local disk of each computer in the cluster. When a mapper completes writing its sorted output files, the MapReduce scheduler notifies the reducers that the output files from that mapper are ready for retrieval. The reducers connect to each mapper and get the files containing sorted (key, value) pairs for their partitions. After the data movement, each reducer takes the files from the mappers, merges them together, preserving the sort order, and starts reducing on the merged input.

3.2.4   *Uses and performance considerations*

Distributed computing systems have been widely employed for data processing for many years. These systems perform well with compute-intensive jobs; however, they require a large amount of network bandwidth to manage massive volumes of distributed data. To mitigate this issue, MapReduce incorporates a data locality feature to maximize performance and save energy by locating compute tasks near the input data (e.g., same node, same rack). This allows for reducing the network bandwidth bottleneck that frequently plagues distributed data analysis systems.

Moreover, since data are evaluated at processing time, MapReduce can be used to handle semi-structured or unstructured data in parallel, as opposed to RDBMS, which are suited for storing and processing structured data. To process those very large volumes of semi-structured or unstructured data in distributed/parallel systems employing hundreds or thousands of computers, the model must tolerate machine failures. When a worker fails, the task is re-executed by another worker. Only in the worst case scenario, when the compute node hosting the master fails, it is necessary to restart the entire MapReduce job.

Thanks to the features described above, MapReduce is commonly used to implement scalable data analysis algorithms and applications that run on numerous machines to efficiently analyze large volumes of data. It was designed to be utilized in a variety of domains, such as data mining and machine learning, social media analysis, financial analysis, image retrieval and processing, scientific simulation, website crawling, machine translation, and bioinformatics. MapReduce allows full exploitation of data parallelism, enabling the efficient execution of such applications in distributed systems, whose complexity is mostly related to the large volume of data to be processed. MapReduce is now widely regarded as one of the most important parallel programming models for distributed systems.

It is worth noting that MapReduce programs are not always guaranteed to be efficient. The fundamental advantage of this programming model is that it only requires writing the map and reduce phases of the program. Nevertheless, in practice, the shuffle phase and the amount of data generated by the map function can significantly impact performance and scalability. Additional modules, such as the combiner, can help reduce the amount of data written to disk and transferred across the network. *Speculative execution* is another common optimization used in MapReduce-based systems. If a node is accessible but underperforming (the so-called *straggler condition*), MapReduce runs a speculative (or backup) duplicate of its tasks on another computer to complete the computation faster.

Furthermore, utilizing the MapReduce paradigm for programs that need to fit data into the main memory of a single machine or a small cluster is typically not an effective option. In fact, MapReduce frameworks are meant to be fault-tolerant and recover from node failures during computation. This is achieved by storing intermediate results in distributed storage. However, this recovery is costly and becomes necessary only when the computation involves a large number of machines and a long computation time. When a failure occurs, a job that requires relatively little time can simply be restarted. Frameworks that keep all data in memory and restart computations when faults occur are faster than MapReduce in such cases.

3.3 The Workflow Model

Workflows have emerged as a useful programming model for dealing with the complexities of scientific and business applications. The widespread availability of HPC systems has enabled scientists and engineers to develop more complex programs to access and handle enormous data repositories on distributed computing platforms. The majority of these applications are intended as workflows that combine data analysis, scientific computation, and complex simulation methods (Belcastro *et al.*, 2019).

A workflow is a series of activities, events, or tasks that must be completed in order to achieve a goal and/or a result. The Workflow Management Coalition[1] defines a workflow as *the automation of a business process, in whole or in part, during which documents, information, or tasks are passed from one participant to another for action, according to a set of procedural rules.* The same description may be used for scientific processes that are made up of numerous processing steps linked together to describe data and/or control dependencies. Here, the term *process* refers to a set of tasks that are connected together with the objective of producing a product, calculating a result, or delivering a service. Hence, each task

---

[1]https://wfmc.org/.

(or activity) is a piece of work that represents one logical step in the entire process.

Workflows, as a programming model, represent well-defined and possibly repeatable patterns or systematic groupings of activities aimed at achieving a certain data transformation (Talia and Trunfio, 2012). They take a declarative approach to express the high-level logic of many types of applications while obscuring low-level details that are not essential for application design. One significant advantage of workflows is that, once defined, they can be stored and retrieved for change and/or re-execution. This allows users to design common patterns and reuse them in multiple contexts. A workflow management system (WMS) facilitates the definition, development, and execution of processes. The coordination of operations of the different activities (or enactment) that compose the workflow is a critical feature of a WMS during workflow execution.

A workflow is programmed as a graph, which consists of a finite set of edges and vertices, with each edge directed from one vertex to another. In a workflow model, vertices represent specific tasks, activities, or stages within the overall process, while edges represent the flow or sequence of tasks, indicating the order in which tasks should be executed. A data analysis workflow, for example, can be designed as a series of pre-processing, analysis, and post-processing tasks. A workflow can be implemented as a software program using a programming language, a library, or a system that allows for the expression of the basic workflow steps and includes mechanisms to orchestrate them.

3.3.1   *Workflow patterns*

Workflow tasks can be combined in a variety of ways (e.g., as sequences or parallel constructs), which allows designers to address the needs of a wide range of application scenarios (see Figure 3.3). In the following sections, the main workflow patterns are discussed.

3.3.1.1   *Sequence*

The *sequence* pattern is one of the simplest and most common patterns for composing tasks in a workflow. It denotes a sequence of tasks that must be completed in a specific order. These tasks are connected by directed edges that indicate the direction of control flow, specifying the sequential order in which the tasks execute.

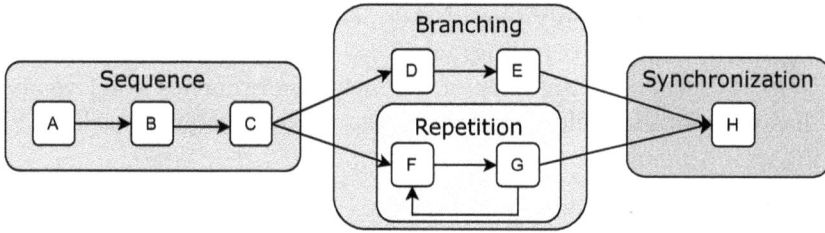

Fig. 3.3. Workflow patterns.

3.3.1.2 *Branching*

The *branching* pattern describes situations in which a branch in a workflow is split into two or more different branches. There are three distinct branching variants that can be identified (Van Der Aalst *et al.*, 2003):

- *AND-split*, where the control flow in a branch splits into concurrent execution flows in each subsequent branch.
- *XOR-split*, where the control flow in a branch is directed into only one of the branches generated from the split construct. The decision on which of the subsequent branches to pursue is based on an evaluation of the conditions associated with each of the outgoing edges.
- *OR-split*, where the control flow in a branch splits into concurrent execution flows in one or more subsequent branches based, as in XOR-split, on the evaluation of conditions associated with each of the outgoing edges. Unlike XOR-split, however, multiple branches can be chosen.

3.3.1.3 *Synchronization*

Synchronization patterns describe situations in which multiple control flows on one or more branches must be merged into a single branch. Such scenarios are common in real-world workflows, where the execution of a specific task must wait for one or more preceding tasks to be completed. Three types of synchronization pattern can occur in practice:

- *AND-join*, where all incoming branches are active and the workflow requires all of them to complete before proceeding to the subsequent task.

- *XOR-join*, where only one of the incoming branches needs to be complete before proceeding.
- *OR-join*, where at least one of the incoming branches is active and has to complete before control is passed to subsequent task.

3.3.1.4   *Repetition*

Repetition patterns describe various ways of specifying repetition. The *arbitrary cycle* pattern, which indicates one or more tasks being repeated within a workflow, is the most general form of repetition. These cycles have an unstructured form and are represented in the model by circular paths rather than by a specific construct. Furthermore, the cycle may have more than one entry or exit point, and not all tasks may be executed the same number of times. In contrast to this type of repetition, the *structured loop* pattern describes a specific set of tasks that are repeated with a specific termination condition, which is evaluated either before or after each iteration. The loop has only one entry and exit point, and it is possible to represent the loop as a specific construct in the workflow. The structured loop is equivalent to implementing repetition with a *while ... do* or *repeat ... until* loop, whereas arbitrary cycles are equivalent to the *goto* statement. A third approach to repetition is the *recursion* pattern, which describes situations in which a specific task is repeated through self-invocation.

3.3.2   *Directed acyclic graphs*

When there are no cycles in a workflow, it is referred to as a directed acyclic graph (DAG), which is the most commonly used programming structure in workflow management (Talia, 2013). In particular, a DAG is:

- *Directed*: If multiple tasks exist, each must have at least one previous or subsequent task, or both. However, some DAGs have multiple parallel tasks, implying that there are no dependencies.
- *Acyclic*: Tasks cannot generate data that reference themselves, potentially resulting in an infinite loop. This means that DAGs do not have any cycles.

The DAG paradigm is effective for modeling complex data analysis processes, such as data mining applications, which can be

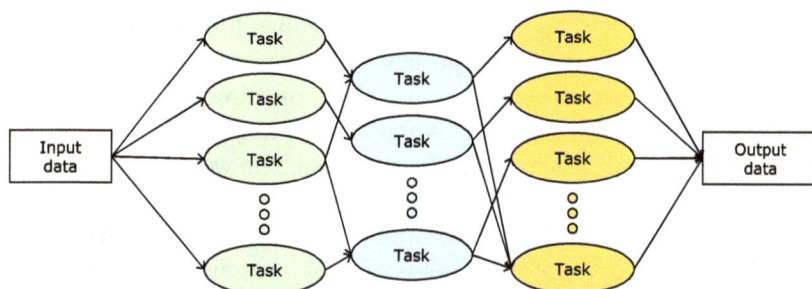

Fig. 3.4. DAG execution flow.

efficiently executed on distributed computing systems, such as a cloud platform. There are two types of dependencies to consider: *data* dependencies and *control* dependencies. Specifically, in the former, the output of a task serves as the input for the next tasks, while in the latter, certain tasks must be completed before starting another task or set of tasks. DAGs can easily model many different types of applications, in which the input, output, and tasks of one application are dependent on the tasks of another (see Figure 3.4). The tasks of a DAG application and their dependencies can be defined in one of two ways:

- *Explicitly*, when dependencies among tasks are defined through explicit instructions (e.g., T_2 depends on T_1).
- *Implicitly*, when the system automatically infers dependencies among tasks (e.g., T_2 reads the input O_1, which is an output of T_1) by analyzing their input–output relationships.

It is worth noting that the DAG model is a strict generalization of the MapReduce model. According to the MapReduce model, distributed computation on a large dataset can be performed using two types of computation phases: map and reduce. One map and reduce pair performs one level of aggregation on the data; however, complex computations typically necessitate multiple such phases, resulting in a DAG of operations. DAGs have proven to be extremely useful in a variety of big data frameworks, including Spark. Some of the benefits of using the DAG model include greater flexibility than MapReduce and better global optimization. In fact, a DAG can be optimized by rearranging and combining operators wherever possible. For example, if we consider two operations, such as *map* and *filter*, an optimization could be to run them in reverse order because filtering may reduce the number of records that experience map operations.

A more complex workflow model relies on directed cyclic graphs (DCGs), where cycles represent some form of implicit or explicit loop or iteration control mechanism. In this case, the workflow graph frequently describes a network of tasks, with nodes representing either services, instances of software component instances, or more abstract control objects. The graph edges represent messages, data streams, or pipes that allow services and components to exchange work or information.

3.4   The Message-Passing Model

The message-passing model is a well-known paradigm that provides the fundamental mechanisms for inter-process communication (IPC) in distributed computing systems, with each processing element having its own private memory. When it comes to IPC, i.e., the mechanism provided by the operating system that allows processes to communicate with one another, processes can communicate with one another via two mechanisms: shared memory and distributed memory or message passing. As a result, parallel programming models are generally classified based on how memory is used.

3.4.1   *From shared memory to message passing*

Each process in a shared memory model has access to a shared address space. Communication between processes via shared memory necessitates the sharing of some variables, which is entirely dependent on how the programmer implements it. Let us suppose we have two processes, P_1 and P_2, that are running concurrently and sharing resources or using information from another process. P_1 generates information about specific computations or resources in use and stores it in shared memory as a record. When P_2 needs to use the shared information, it will inspect the shared memory record, take note of the information generated by P_1, and act accordingly. Processes can use shared memory to extract information as a record from another process as well as for other purposes. In addition, shared memory can be used by processes in order to extract information from another process as a record and to deliver any specific information to other processes.

In the message-passing model, on the other hand, an application runs as a collection of autonomous processes, each with its own local memory and communicating with other processes via the sending and receiving of messages. When data are sent in a message, the sending and receiving processes must collaborate to transfer the data from one's local memory to the other's local memory. If the two processes, P_1 and P_2, in the preceding example want to communicate with each other, they must first establish a communication link and begin exchanging messages using basic primitives. This abstraction can be implemented in a variety of ways. Although the message-passing concept appears simple, it necessitates a number of design decisions which determine many different message-passing implementations in practice. More details about message basic principles and primitives are given in the following sections.

A comparison between the two models is illustrated in Figure 3.5.

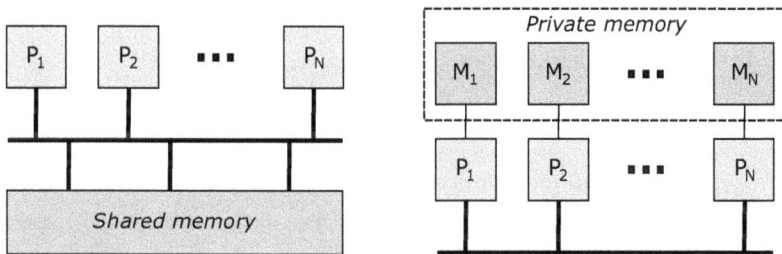

Fig. 3.5. Shared memory vs. message passing.

3.4.2 *Message-passing primitives*

There are two main primitives of the message-passing model, which are used by a set of processes to exchange messages during their execution (see Figure 3.6):

- *Send(destination, message)*: A process sends a *message* to another process identified as the *destination*.
- *Receive(source, message)*: A process receives a *message* from another process identified as the *source*.

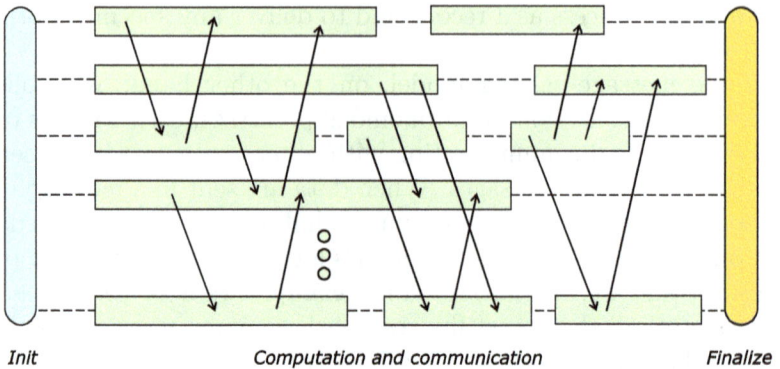

Init   Computation and communication   Finalize

Fig. 3.6.   Message-passing execution flow.

The method of implementing the sending and receiving operations determines different message-passing implementations in practice. Specifically, we can distinguish between *direct* or *indirect*, *buffered* or *unbuffered*, and *blocking* or *non-blocking* message passing.

3.4.2.1   *Direct and indirect message passing*

In direct message passing, there is a link between two processes that exchange data, i.e., the receiver's identity is known and a message is directly sent. One significant disadvantage of direct message processing is its lack of modularity. In fact, changing a process' identity requires updating every sender and receiver with a connection to the process.

On the other hand, in indirect message passing, mailboxes or ports are exploited for message delivery, which can be bound to a receiving process. It differs from direct message passing in that the same port can later be assigned to another process. A sender has no idea which process will receive its message. Furthermore, multiple processes can send messages to the same port, allowing for multi-process links and greater flexibility. In indirect message passing, the creation of a mailbox comes before the send and receive primitives:

(1) $A = createMailbox()$: create a mailbox and assign it to A.
(2) $Send(A, message)$: send a message to mailbox A.
(3) $Receive(A)$: receive a message from mailbox A.

3.4.2.2 *Blocking and non-blocking message passing*

Another significant distinction is between *blocking* and *non-blocking* message passing. The terms blocking/non-blocking and synchronous/ asynchronous are used interchangeably. A *send* operation is blocking if the sender must wait for the receiver to receive a message. Similarly, if the receiver must wait until the message is received, the *receive* operation is blocking. As a result, the sending process creates the message containing the data to be shared with the receiving process, including a header indicating which processing element and process the data should be sent to, and transmits it to the network. Once the message has been inserted into the receiver's buffer, the sending process continues its execution. The receiving process must be aware that it is expecting data, and it indicates its readiness to receive a message by executing a *receive* operation. As a kind of synchronization, if the expected data has not yet arrived, the receiving process will block (or wait) until it does arrive.

On the other hand, non-blocking is deemed asynchronous. The sender delivers the message and continues its operations, and the receiver gets either a valid or a null message. How the receiving process knows that the message has arrived is a key issue in a non-blocking receive primitive. It is worth mentioning that for a sender, it is more natural to act in a non-blocking manner during message passing since the message may need to be sent to several processes. However, the sender expects acknowledgment from the receiver in case the send fails. Similarly, since the information from the received message may be used for subsequent execution, it is more reasonable for a receiver to block. Similarly, if the message transmission continues to fail, the receiver will be forced to wait indefinitely. For these reasons, there are three fundamentally recommended combinations:

- Blocking send and blocking receive, called a *rendez-vous* communication.
- Non-blocking send and non-blocking receive.
- Non-blocking send and blocking receive, which is the most used.

3.4.2.3 *Buffering*

One distinction between message-passing models can be made by considering the size of the receiver's queue. There are three alternatives available:

- *Zero capacity queue (no queue)*: The sender must wait for the receiver to be ready to receive the message. This enforces a rendez-vous.
- *Bounded queue*: The queue can only hold a maximum of n messages, or n message bytes. Messages can be queued as long as the queue is not full; otherwise, the sender will be forced to block.
- *Unbounded queue*: Senders are never required to wait. Designers should use caution in this decision since physical resources are limited, and having too many messages in the queue might have dangerous consequences.

3.4.3   *Group communication*

One-to-one communication (also known as point-to-point or unicast communication) is the most basic kind of message-based communication, in which a single-sender process sends a message to a single-receiver process. Several highly parallel distributed applications require a message-passing system to additionally include group communication primitives for performance and simplicity of development. There are three forms of group communication conceivable, depending on whether there are single or multiple senders and receivers: one-to-many, many-to-one, and many-to-many.

3.4.3.1   *One-to-many communication*

In one-to-many communication, there are multiple receivers for a message sent by a single sender. *Multicast* communication is another name for one-to-many communication. Specifically, message receiver processes create a group that can be *closed* or *open*. A closed group is one in which only the group's members can send messages to the group, while an outside process cannot send a message to the entire group but only to a specific member. On the other hand, an open group is one in which any process in the system can send a message to the group as a whole. A special case of multicast communication is *broadcast* communication, in which the message is sent to all processors connected to a network.

3.4.3.2 *Many-to-one communication*

In many-to-one communication, multiple senders send messages to a single receiver. The single receiver may be *selective* or *non-selective*. A selective receiver identifies a single sender; a message exchange occurs only when that sender sends a message. A non-selective receiver defines a set of senders, and if one of those senders sends a message to this receiver, a message exchange occurs. Non-determinism is a key issue in many-to-one communication schemes since it is unknown which member (or members) of the group will have their information available first.

3.4.3.3 *Many-to-many communication*

In many-to-many communication, multiple senders send messages to multiple receivers. The issue of ordered message delivery is critical in many-to-many communication schemes. Ordered message delivery guarantees that all messages are delivered in an order acceptable to the applications sent to all receivers.

3.5 The BSP Model

Many graph and linear algebra algorithms require several iterations, in which the output of one iteration is fed as the input of the next one. This is not only slow but also inefficient because data must be written to files at the conclusion of each iteration and moved between compute nodes. For example, for each phase of a graph processing algorithm, it could be necessary to broadcast the full state of the graph over the network, which entails significant additional communication costs. The bulk synchronous parallel (BSP) (Valiant, 1990) is a parallel computation model developed by Leslie Valiant of Harvard University. In the late 1980s, designing computers that could handle parallel programming grew increasingly challenging due to the lack of a single reference model. Valiant advocated for creating a paradigm that connected hardware and software in the same way as Von Neumann's model did but for parallel machines. The Valiant paradigm allows the programmer to avoid costly memory and communication management while achieving a low degree of synchronization.

3.5.1   *The superstep*

A computer based on the BSP model consists of the following components (see Figure 3.7):

(1) A number of processing elements (PEs), each performing local computations.
(2) A router that delivers messages point-to-point between pairs of processing elements.
(3) A hardware feature that allows all or a portion of the processing elements to be synchronized at regular intervals of L time units, where L is the communication latency or synchronization periodicity parameter.

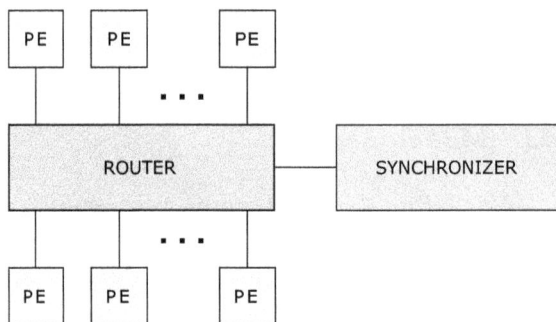

Fig. 3.7.   Architecture of a BSP machine.

A computation is made up of a series of *supersteps*. Each processing element is assigned a task in each superstep that consists of some combination of local computing steps, message broadcasts, and message arrivals from other processing elements. A global check is performed after each period of L time units to check whether the superstep has been completed by all processing elements. If so, the machine proceeds to the next superstep. In each superstep, the following operations are performed (see Figure 3.8):

(1) *Concurrent computation*: Each processor performs computations using local data asynchronously.

(2) *Global communication*: The processes exchange data among themselves in response to requests made during local computation.

(3) *Barrier synchronization*: When a process reaches the barrier, it expects all other processes to have reached the same barrier.

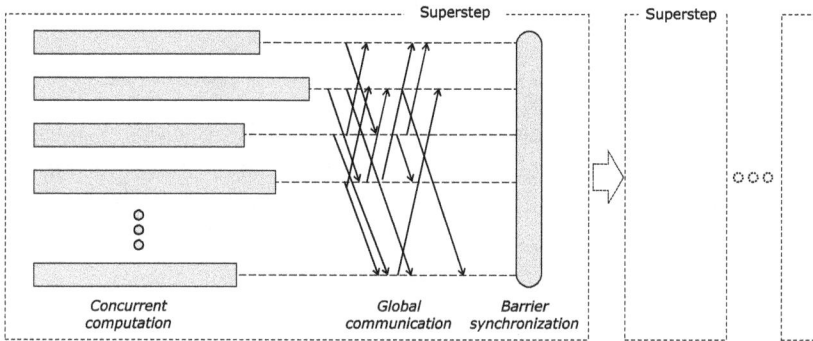

Fig. 3.8.   BSP execution flow.

Communication and synchronization are totally decoupled, ensuring that all processes in a superstep are independent. Furthermore, this technique avoids issues caused by synchronous message passing across processes (e.g., deadlocks). Because of these characteristics, the BSP programming style is a particularly reliable approach for designing scalable parallel systems.

3.5.1.1   *Communication*

Due to the various simultaneous operations in a parallel program and the complexity of the connections, communication management in many parallel programming systems is relatively onerous. The BSP model, on the other hand, considers communication actions *en masse*. This has the effect of putting a time limit on how long it takes to send a batch of data. It also treats all superstep communication actions as a single unit and assumes that all individual messages delivered as part of that unit have a constant size. Let h be the maximum number of inbound or outgoing messages for a specific superstep and g be the ability of the communication network to transfer data, then the time necessary for a processor to send h messages of size one in a superstep can be derived as hg. The value g is called the communication throughput ratio, or *gap*, in the sense of bandwidth inefficiency. A message of length m will certainly take longer to send than a message of unit size. The BSP model, on the other hand, does not discriminate between a message of length m and m messages of length one. In such circumstances, it is straightforward to deduce that the cost is mg. In practice, each parallel computer's estimate of the parameter g is obtained empirically.

3.5.1.2 *Synchronization*

The communication that characterizes the BSP paradigm necessitates synchronization via barriers. Although barriers are potentially costly, they eliminate the risk of deadlocks or livelocks since they cannot build circular dependencies between data. However, various issues influence the cost of synchronizing the barriers:

- The cost imposed by a difference in the completion times of different local computation steps. Consider the following scenario: All but one of the processes have finished their work for a particular superstep and are waiting for the last process, which still has a lot of work to perform. To address this issue, each process might be assigned a task proportionate to the others.
- The cost of ensuring global consistency across all processors. This is heavily dependent on not only the communication network but also the availability of special-purpose hardware for synchronization and the way in which processors handle interrupts.

3.5.2 *Cost of a BSP algorithm*

All the parameters discussed above, i.e., h, g, and L, are related to each other. If L is equal to the time interval necessary for the communication and computation of a superstep and hg indicates the time taken by each processor to send h messages in a superstep, then we must ensure the validity of the relationship $L \geq hg$. In this way, a lower limit is set on the value of L: The periodicity must be established in such a way as to guarantee at least that h messages are exchanged in a superstep considering the time of each processor, i.e., hg. There is also a relationship between the parameters g and p (the total number of processors): As the value of p increases, communication is affected, as there will be more parallel tasks that will have to communicate with each other. It is therefore important to keep g as low as possible so as not to excessively increase the communication time and to ensure the efficiency of the model. The total cost of a superstep s can be expressed as $T_s = w_s + h_s g + L$, where w indicates the total cost of the computation that occurred in the superstep. Note that processors with less than w FLOPS have to wait, and this waiting time is called idle time. Therefore, given S the total number of supersteps, the total cost of a BSP algorithm is given by

$$T = \sum_{1 \leq s \leq S} T_s = \sum_{1 \leq s \leq S} (w_s + h_s g + L) = W + Hg + SL$$

where W and H are the total cost of computation and communication respectively.

3.5.3 *Shared memory-based BSP model*

The BSP model does not directly support shared memory, broadcasting, or combining. However, these features can be achieved by emulating a parallel random access machine (PRAM) on a BSP computer. A PRAM is made up of an infinite number of processors, each of which is connected to a shared memory unit with an infinite capacity. Processors can only communicate with one another via shared memory, to which they are linked by a memory access unit (MAU), while the calculation is entirely synchronous. Several PRAM variations have been introduced, as depicted in Figure 3.9.

		WRITE	
		exclusive	concurrent
READ	exclusive	**EREW** *a memory cell can be accessed by not more than one processor at a time*	**ERCW** *a memory cell can be written by more than one processor at a time*
	concurrent	**CREW** *a memory cell can be read by more than one processor at a time*	**CRCW** *a memory cell can be both read and written by more than one processor at a time*

Fig. 3.9. Classical variations of a parallel random access machine.

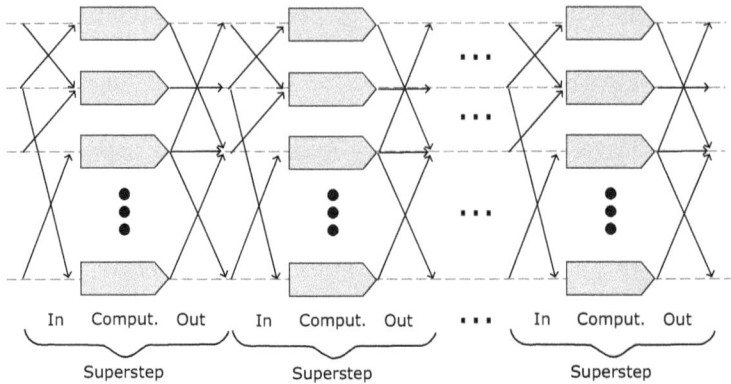

Fig. 3.10.   BSPRAM execution flow.

Tiskin (1998) presented a PRAM-based variation of the BSP machine, designed to facilitate shared-memory-style BSP programming by utilizing local memory for individual processors and shared global memory. A bulk-synchronous PPRAM (BSPRAM) is formally composed of p processors with fast local memory and a single shared main memory. The calculation is done in supersteps, just like in BSP. A superstep comprises three phases: input, local computation, and output, during which processors read data from and write data to the main memory. Between supersteps, the processors are synchronized, while the computation within a superstep is asynchronous. Figure 3.10 depicts the BSPRAM execution flow.

Similarly to PRAM, concurrent access to main memory in a single superstep might be permitted or denied. For example, in an exclusive read exclusive write BSPRAM (EREW BSPRAM), each cell of the main memory may be read from and written to only once in each superstep, while in a concurrent read concurrent write BSPRAM (CRCW BSPRAM), concurrent access to the main memory is not restricted.

3.6   The SQL-Like Model

With the exponential growth of data to be stored in distributed network settings, relational databases exhibit scalability restrictions that dramatically affect querying and analytical efficiency

(Abramova *et al.*, 2014). Relational databases cannot scale horizontally across multiple computers, making it difficult to store and manage the massive volume of data generated every day by numerous applications. In recent years, the "not only SQL" (NoSQL), or non-relational database architecture, has gained popularity as an alternative or complement to relational databases in order to achieve horizontal scalability of basic read/write database operations deployed over multiple servers (Cattell, 2011).

3.6.1 *From NoSQL to SQL-like model*

NoSQL databases address various challenges of big data storage and management, providing horizontal scaling of continuous read/ write operations distributed over multiple servers. Instead of the atomicity, consistency, isolation, and durability (ACID) model used by relational databases, NoSQL databases typically adhere to the basic availability, soft-state, and eventual consistency (BASE) model, which removes the requirement of consistency after every transaction to support the processing of multiple instances on different servers at the same time (see Section 2.3.7).

Although NoSQL systems enable the effective processing of massive amounts of fast-moving data, many applications, such as the handling of financial transactions or personal data (e.g., health information), still require ACID compliance for user protection and privacy. As a result, relational databases, such as Oracle, MySQL, Microsoft SQL Server, and PostgreSQL, continue to be more popular than the most popular NoSQL options, such as MongoDB, Redis, and Cassandra. Furthermore, NoSQL databases are frequently unsuitable for data analytics. For these reasons, much effort has been expended in recent years on designing MapReduce systems to query and analyze relational data more efficiently.

SQL-like systems attempt to combine the efficiency and query capabilities of MapReduce programming with the simplicity of the SQL-like language in order to enable the construction of simple and efficient data analysis applications. In fact, while MapReduce programming can address scalability issues and reduce querying times, low-skilled individuals may find it difficult, as it requires writing complex map and reduce functions for simple tasks (e.g., sum or

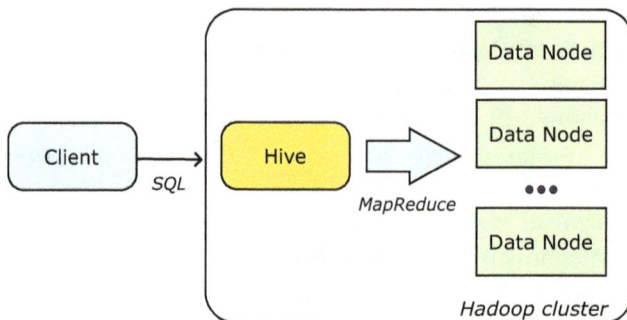

Fig. 3.11.   Developing SQL on big data engines.

average calculation, row selections, or counts), resulting in a significant waste of time (and money) for businesses. To address this issue, several solutions (e.g., Apache Hive) have been created to improve the query capabilities of MapReduce systems and to make the development of basic data analysis applications using a SQL-like language easier (see Figure 3.11).

3.6.2   *Why use SQL on big data?*

SQL has become the *de facto* tool for developers, database managers, and data scientists to access and manipulate data. It is used in a variety of commercial products and applications for data querying, modification, and visualization. The primary benefits of employing a SQL-like paradigm can be described as follows:

- *Declarative language*: SQL is a declarative language, which means it describes which parts of the data are undergoing transformations and operations. Users can easily comprehend the semantics of SQL queries by reading them, just as they can understand literal descriptions. SQL statements are often classified into three types depending on their use:

  - *Data Definition Language (DDL)*: to create or modify the structure/constraints of tables and other objects in a database, including the creation of schemes and databases.
  - *Data Manipulation Language (DML)*: to insert, delete, or update the data in the existing structures/objects in database tables.

– *Data Querying Language (DQL)*: to extract/retrieve and work with the data in a database without modifying data.

- *Interoperability*: Since SQL is a standardized language, each system can provide its own implementation while maintaining compatibility between platforms and frameworks. Despite the different SQL variations among big data processing frameworks, such as Hadoop Query Language (HQL) in Hive and Cassandra Query Language (CQL) in Cassandra, users can still easily understand the syntax used for writing programs.

- *Data-driven*: All operations and primitives in SQL reflect transformations and modifications of the input dataset (tables of data in SQL). As a result, SQL is one of the most convenient programming models for data-centric applications in both traditional databases and recent big data environments.

For these reasons, SQL-like systems are often used to overcome the complexity of writing MapReduce programs, even for simple operations (such as row aggregations, selects, or counts), while retaining query speeds and scalability. These systems use MapReduce to automatically apply some optimizations to the full query. Data manipulation, data querying, and reporting on massive repositories are their primary application domains. In particular, data must be cleansed and processed before being imported into a destination suitable for analytics, e.g., a distributed file system or data warehouse. For this purpose, extract, transform, and load (ETL) processes are utilized. A SQL-like model allows one to read unstructured data, analyze them as needed, and then load them into a target datastore for decision support systems using a series of SQL-like queries.

To summarize, in the following, we highlight some of the primary goals of SQL-like model for big data processing. Starting with SQL databases, such as Microsoft SQL Server, MySQL, and PostgreSQL, the major challenge with querying big data is that querying operations cannot be scaled up without incurring significant costs. As a result, SQL must be supported by distributed architectures in order to scale up data storage and compute across machine clusters. Another purpose of designing a SQL solution for big data is to avoid moving data from the datastore (such as a distributed file system) to an external store for analytics. The SQL engine could work with the data stored in the data node, completing the computation at a cheaper cost. Furthermore, because data are immediately accessed on the storage cluster, combining SQL and big data results in instant availability of ingested data. This is known as a *query-in-place* technique, and its advantages include low-latency *ad hoc* SQL queries on massive datasets. Since there is no need to maintain a separate analytic database, data movement from one system to another is reduced. All of these factors result in lower operating costs and greater overall efficiency in data processing.

3.6.3   *Data partitioning*

A crucial aspect of using SQL statements to query big data is data partitioning. Partitioning divides data in a table into one or more columns based on the values of the partitioned columns, generating distinct files and/or directories. A simple SQL query, in general, reads the entire table, resulting in a time-consuming operation. Instead, when the table is queried using a partitioning strategy, only the required partitions are scanned, which lowers I/O costs by avoiding reading data that is known not to meet the query. Because the data are partitioned across directories, it is faster for a query to process the partitioned part of the data than a full scan. However, having too many partitions generates a huge number of files and directories in the distributed file system, which is costly for the master node because it must retain all metadata in memory.

3.7   The PGAS Model

The partitioned global address space (PGAS) (De Wael *et al.*, 2015) paradigm is a parallel programming model that attempts to enhance programmer productivity while still achieving high performance. The essential concept of PGAS is that, while a globally shared address space enhances productivity, a separation between local and distant data accesses is necessary to facilitate performance improvements and support scalability on large-scale parallel architectures. For that purpose, PGAS retains a global address space. A PGAS program consists of several processes running the same code on various nodes at the same time. Each process has a rank, which is the node index on which it runs. Each process has access to a global address space that

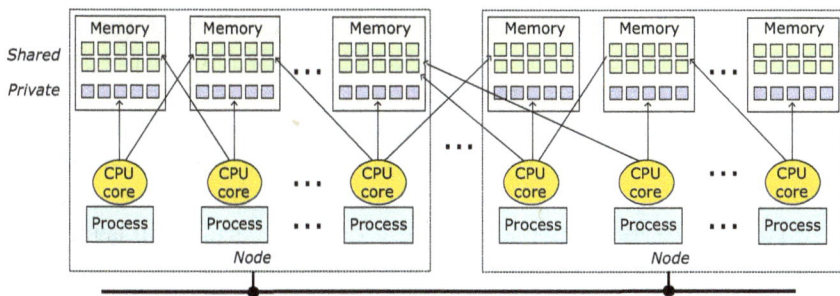

Fig. 3.12. Example of the PGAS model.

is partitioned into local address spaces. Local addresses are directly accessible, whereas remote addresses belonging to distinct processes are accessed via API calls. An example of the PGAS model is illustrated in Figure 3.12.

Languages based on the PGAS paradigm treat the address space as a global environment. This implies that a thread or a process gets a pointer to data that can be located anywhere in the system and can also read or write remote data that are local to other threads. Every PGAS language distinguishes two memory areas in the address space: shared memory, which includes data available to all threads, and private memory, which contains data only accessible to the thread that owns that area. Each thread has its own part of private space as well as a section of shared space.

The properties of PGAS languages are placed in the middle of the message-passing and shared memory models. In fact, it combines the data locality (partitioning) features of message passing with the programmability and data referencing simplicity of a shared-memory (global address space) model.

3.7.1 *Parallelism in PGAS*

Based on the PGAS model, we distinguish three major parallel execution models:

- *Single program multiple data (SPMD)*: A fixed number of threads are spawned at program startup, each of which executes the same program.
- *Asynchronous PGAS (APGAS)*: At program startup, a single thread begins execution at the program's entry point. Constructs

are provided to dynamically spawn new threads that run in the same or remote address space partitions, where each spawned thread may execute a different code.

• *Implicit parallelism*: The code contains no visible parallelism or parallelism control directives, implying that the program describes a single thread of control. Multiple control threads may be spawned at runtime to speed up computation, but this parallelism is implicit in the program's code.

3.7.2   Memory and cost function

The PGAS model divides memory space into *places*. A place is a computational node with which a specific process or thread is associated. The thread associated with a location can access the memory locations of that place at a low and uniform cost. All other memories associated with other places and thus with other threads are remote and more expensive to access. The actual cost varies depending on the underlying hardware. The relationship among the different places is defined on the basis of the interconnection topology that binds them together. The most common topology is that in which each place is identified by a numerical index within the range $[0, n]$, where n is the total number of places. The hierarchical tree is another common topology.

The presence of the non-uniform memory access (NUMA) shared-memory model underlying the PGAS model is highlighted by the cost function, which characterizes the memory accesses. A PGAS language typically employs a two-level cost: cheap and expensive. Memory locations close to the source of the access request are cheap, while distant memory locations are expensive. As a result, two elements contribute to the cost calculation: the place from which the access request originates and the place of the request's destination, where the data of interest are located.

3.7.3   Data distribution

The distribution of data across places can be used to categorize PGAS languages. When programmers can specify the distribution of data, we refer to an explicit model; otherwise, we refer to an implicit model. Three common distribution models can be identified:

- *Cyclic*, in which data are partitioned in consecutive chunks that are cyclically arranged into places.
- *Block*, in which data are partitioned into equally-sized and consecutive chunks that are distributed in different places.
- *Block-cyclic*, in which data are partitioned into parameterized-size chunks that are sequentially arranged in different places in a cyclic manner.

When dealing with multidimensional data (e.g., matrices), these three distribution models can be used in a variety of ways. A distribution model, for example, can be applied per dimension. Another strategy is to use a distribution for each dimension while ignoring one of them (for example, the column dimension). Otherwise, a distribution model can be applied to flattened data.

3.8 Models for Exascale Systems

The adoption of exascale systems offers a great opportunity; however, their design and implementation are quite complex due to a number of challenges, as discussed in Section 2.7.1. Scalability, network latency, reliability, reproducibility, and robustness of procedures and operations accessible to developers for transferring and managing data are the main challenges in designing applications on exascale systems. Indeed, processing very large data volumes requires operations and new algorithms capable of scaling in loading, storing, and processing massive amounts of data, which must typically be partitioned into very small data grains and analyzed through thousands or even millions of simple parallel operations.

3.8.1 *The role of programming models in exascale systems*

Several reports from different organizations have identified a number of obstacles to exascale computing. These difficulties include issues with performance, scalability, and productivity, in addition to the more recent concerns of energy efficiency and robustness. Recent HPC architectures demonstrate how cutting-edge technologies, such as graphics processors, system-on-a-chip architectures, and non-volatile memory, can provide innovative solutions to some of these issues. Although early use of these systems demonstrated performance and power advantages, portability and performance stability issues remain. These issues, taken together, are preventing the scientific community from adopting these solutions.

Supercomputers have become an essential tool in a wide range of scientific fields, including quantum physics, weather forecasting, climate research, molecular modeling, and physical simulations. To enable future scientific discoveries, effective parallel applications capable of meeting processing demands must be developed. Modern HPC systems are made up of hundreds of thousands of processing nodes, whose increasing sizes prevent programmers from having a complete view of the system. Because of this, the programmability of an HPC system substantially influences its total performance; therefore, programming models should provide a high level of scalability to aid the creation of next-generation exascale supercomputers.

Programmers must also have an abstraction that allows them to manage hundreds of millions to billions of concurrent threads, enabling them to structure programs into comprehensible pieces, which is critical for system clarity, maintenance, and scalability. It also enables increased programmability by building new parallel programming languages on top of existing ones or even completely from scratch. This makes abstraction an important part of most parallel paradigms and runtimes.

3.8.2 Requirements of exascale models

The trade-off between sharing data among processing components and local computation is one of the most significant aspects to consider in applications that run on exascale systems and analyze big data. A scalable exascale programming model must include at least the following mechanisms (Talia, 2019):

- Parallel data access, which increases data access bandwidth by splitting data into several chunks and accessing different data elements concurrently;
- Fault resiliency to deal with the possibility of one of the communication sides failing during non-local communication;
- Data-driven local communication to reduce data exchange;

- Data processing on limited groups of cores to concentrate computation on localities of exascale machines;
- Near-data synchronization to reduce the overhead generated by synchronization among numerous distant cores;
- In-memory analytics which reduces reaction time by caching data in the RAMs of processing nodes;
- Locality-based data selection to reduce latency by keeping a subset of data required locally available.

3.8.3 *Limitations of current programming models*

Reaching exascale in terms of computing nodes requires a shift from the current control of thousands to that of billions of threads, as well as an adaptation of current models to cope with an increasing level of failures. According to Gropp and Snir (2013), the five most essential properties of programming models that are affected by the exascale transition are thread scheduling, communication, synchronization, data distribution, and control views. In the following section, the limitations of the main programming models for exascale computing are discussed (Da Costa *et al.*, 2015).

3.8.3.1 *Limitations of message-passing programming model*

The current aim of exascale systems is to use distributed memory parallelism, and hence, the message-passing architecture is likely to be adopted in part. The model's most popular system implementation, MPI, has been proven to run with millions of cores in specific situations. However, MPI is built on standard sequential programming languages and enhanced with low-level message-passing techniques, requiring users to deal with all aspects of parallelization, ranging from data and work distribution to cores, communication, and synchronization. MPI is designed primarily for static data distribution and is hence unsuitable for dealing with dynamic load balancing. Furthermore, it has been demonstrated that the many-to-many communication schemes used in message-passing models are not scalable because the most commonly used implementations frequently assume a fully connected network with dense communication patterns. Other limitations are particularly connected with collective access to the I/O request and data splitting. Indeed, I/O is a bottleneck in MP-based systems, indicating that the current model should be revisited.

3.8.3.2 *Limitations of shared-memory programming models*

Exascale systems are expected to support hundreds of cores on a single CPU or GPU. In the case of medium-sized parallel systems, using shared-memory systems is a viable alternative to message passing because it shifts the parallelization effort from the programmer to the compiler. The most popular shared-memory systems use a parallelism control model that does not allow for data distribution control and employs non-scalable synchronization mechanisms, such as locks or atomic sections. Furthermore, the global view of data encourages joint synchronization of all threads' remote data accesses comparable to the local ones, which leads to inefficient programming.

3.8.3.3 *Limitations of heterogeneous programming*

Because of the benefits of peak performance and energy efficiency, clusters of heterogeneous nodes made of multi-core CPUs and GPUs are increasingly being used for HPC. Application developers frequently use a combination of parallel programming paradigms to fully utilize the computing capabilities of such platforms. However, heterogeneous computing introduces a new challenge in dealing with the variety of execution environments and programming models. The design of single-node hardware is becoming more heterogeneous. Likewise, many of today's greatest HPC systems are clusters of diverse computing device architectures. It is difficult to port single-node multi-device applications to clusters that include heterogeneous compute device architectures, and it also necessitates the use of a communication layer for data transmission between nodes. Because writing programs on such platforms is error-prone, new abstractions, programming models, and tools are necessary to address these issues.

3.8.4 *Models for exascale programming*

To cope with the limitations of existing programming models in exascale systems, new programming models have been recently proposed.

These models exploit or extend the existing ones, taking into account the specific characteristics of exascale environments.

Among them is Legion (Bauer *et al.*, 2012), a distributed memory programming model designed to provide high performance on current parallel architectures with diverse processors and deep memory hierarchies. It is built on the use of *logical regions* to define data organization and provide explicit relationships for reasoning about locality and independence. A logical region identifies a group of objects and can be allocated, removed, and stored in data structures dynamically. Regions can also be supplied as inputs to distinct functions known as *tasks*, which read data in specific regions and provide locality information. Logical regions can be divided into disjointed or aliased (overlapping) subregions, providing information for determining computation independence.

Charm++ (Kalé and Krishnan, 1993) is another distributed memory programming model in which a program specifies collections of interacting objects that are dynamically mapped to processors by the runtime system. Charm++ implements an asynchronous, message-driven, and task-based model with movable objects and an adaptive parallel runtime system that controls execution. It manages communication overlap, load balance, fault tolerance, checkpoints for split execution, and power management. In Charm++, overdecomposition allows the programmer to divide an application into many small objects, each representing a coarse work and/or a data unit. The number of such objects may greatly exceed the number of processors. Moreover, objects can be migrated among processors. In this way, operations may send data to logical objects rather than physical processors.

DCEx (Talia *et al.*, 2019) is a programming model based on the PGAS paradigm for implementing data-centric, large-scale parallel applications on exascale computing platforms. It is built on data-aware basic operations for data-intensive applications, allowing for the scalable usage of a massive number of processing elements. The DCEx model employs private data structures and restricts the amount of data exchanged by concurrent threads. Computation threads use near-data synchronization based on the PGAS paradigm to run close to data. The core idea behind DCEx is to structure programs into data-parallel blocks, which are the memory/storage hierarchy's units of shared- and distributed-memory parallel computation, communication, and migration.

X10 (Charles *et al.*, 2005) is an APGAS-based programming model that introduces locations as a computational context abstraction with a locally synchronous view of shared memory. An X10 computation is distributed among a large number of places, each of which stores some data and performs one or more activities (i.e., lightweight threads of execution) that may be dynamically created. An activity can use one or more memory regions in the place where it resides synchronously.

Chapel (Deitz *et al.*, 2006) is also an APGAS-based programming model that uses high-level language abstractions to facilitate general parallel programming. Chapel provides a global-view programming model (i.e., global-view data structures and a global view of control) that increases the level of abstraction used to define both data and control flow. Global-view data structures are arrays and other data aggregates whose sizes and indices are represented globally, even if their implementations may distribute them across parallel system locales. A locale is an abstraction of a target architecture's unit of uniform memory access, which means that all threads within a locale have similar access times to any single memory address. A global view of control means that a user's application starts with a single logical thread of control and then incorporates parallelism through the use of certain language concepts.

UPC++ (Zheng *et al.*, 2014) is a C++ library that offers classes and methods supporting PGAS programming. UPC++ offers tools for describing dependencies between asynchronous computations and data transfer and for efficient one-sided communication. It also allows moving computation to data via remote procedure calls, enabling the efficient implementation of complicated distributed data structures. Among its main features, UPC++ provides interfaces that are compliant by design with standard C++. Global pointers, RPC-based asynchronous programming, and futures are the three primary programming concepts of UPC++. Global pointers allow the efficient exploitation of data locality, while futures support the development of asynchronous programs, managing the availability of the data resulting from the computation.

As a final remark, algorithms that frequently use all-to-all communication do not scale well in exascale environments. For this reason, hybrid systems have been proposed to cope with scalability issues by using MPI for internode parallelism and a shared-memory programming model for intranodal parallelism. As an example, MPI + X systems (e.g., MPI + OpenMP) enable neighbor collectives in order to provide scalable "all-to-some" communication patterns that limit data transfer only to specific processor areas.

Chapter 4

Tools for Big Data Applications

This chapter describes programming languages, libraries, and tools used for developing scalable big data applications. The programming features and mechanisms of frameworks such as Hadoop, Spark, and Storm are described. For each programming tool, a few real-world examples of big data applications are presented.

4.1 Main Features

Programming systems for big data analysis are characterized by several features that can be used for classification purposes. The most important features are the *level of abstraction* and the *type of parallelism*.

The level of abstraction of a system refers to its programming capabilities to hide the low-level details of a solution (e.g., a function, a data structure, or a communication protocol). A *low level of abstraction* allows programmers to exploit low-level APIs, mechanisms, and instructions that are powerful but not trivial to use. A *medium level of abstraction* allows programmers to define applications by using a limited set of programming constructs, hiding the low-level details that are not fundamental for application design. A *high level of abstraction* allows developers to build applications using high-level interfaces, such as visual IDEs or abstract models, with high-level constructs not related to the running architecture.

The type of parallelism describes the way in which a system expresses parallel operations and how its runtime supports the execution of concurrent operations on multiple nodes or processors. Two main types of parallelism can be achieved: *data parallelism*, in which the same code is executed in parallel on different data elements, and *task parallelism*, in which the different tasks that compose an application are executed in parallel.

In Chapter 5, we discuss more in detail the above features and other aspects that can be used for classifying and comparing systems.

4.2   MapReduce-Based Programming Tools

The most commonly used open-source framework implementing the MapReduce programming model is Apache Hadoop. It is a general-purpose framework designed to manage and process very large amounts of data on parallel and distributed systems, in which the two basic operations of the MapReduce model are executed concurrently. Hadoop enables the development of data-intensive scalable applications using different programming languages. The programming approach used by Hadoop allows developers to abstract from classical distributed computing issues, such as data locality, workload balancing, fault tolerance, and network bandwidth saving.

A few minor implementations of the MapReduce model have been designed and implemented in the past years, such as Phoenix++ and Sailfish; however, none of those have ever achieved success like Hadoop. Phoenix++ is based on C++ and leverages multi-core chips and shared-memory multi-processors. The Phoenix++ runtime automatically handles thread creation, data partitioning, dynamic task scheduling, and fault tolerance. Sailfish is a MapReduce framework for large-scale data processing that exploits batch transmission from mappers to reducers to improve application performance. Sailfish uses an abstraction for supporting data aggregation, called *I-files*, which adapts the original model to efficiently batch data written to and read by multiple nodes.

The following section introduces Apache Hadoop, presents its software architecture, and discusses the parallel execution flow through programming examples.

4.2.1 *Apache Hadoop*

Apache Hadoop[1] is widely used to develop *batch applications*, and over the years, it has been adopted by most of the leading IT companies, such as Yahoo!, IBM, and Amazon. For example, Yahoo! used it for developing advertising systems, web searches, and scaling tests. However, it is suitable only for batch processing, resulting in inefficiency with highly iterative applications that repeatedly perform operations on the same set of data. This is due to disk-based processing on the distributed file system when computing intermediate results with the MapReduce model (Verma *et al.*, 2016). Nevertheless, the project is supported by a large user community, and its diffusion is attributed to wide support for different programming languages and to constant updates and bug fixes by a massive open-source community.

Hadoop provides a *low level of abstraction* because programmers can define an application using APIs that are powerful but not easy to use. In fact, such APIs are built on the computing infrastructure and require a low-level understanding of the system and the execution environment to deal with issues related to distributed file systems, networked computers, and distributed programming (Wadkar *et al.*, 2014). Developing an application using Hadoop requires more lines of code and development effort when compared to systems providing a higher level of abstraction (e.g., Airflow, Pig, or Hive); however, the code is generally more efficient because it can be fully tuned.

Hadoop is designed for exploiting *data parallelism* during map/reduce steps. In fact, input data are partitioned into chunks and processed by different machines in parallel. Data chunks are replicated on different nodes, ensuring high fault tolerance along with checkpoints and recovery. However, the partitioning strategy does not guarantee efficiency when it is needed to access a large number of small files.

In addition to the MapReduce programming model, the Hadoop project includes many other modules, such as:

- *Hadoop Distributed File System (HDFS)*, a distributed file system providing fault tolerance with automatic recovery, portability

---

[1]https://hadoop.apache.org/.

across heterogeneous and low-cost commodity hardware and operating systems, high-throughput access, and data reliability.

- *Yet Another Resource Negotiator (YARN)*, a framework for cluster resource management and job scheduling.
- *Hadoop Common*, which includes utilities and libraries that support the other Hadoop modules.

In particular, thanks to the introduction of YARN in 2013, Hadoop turned from a batch processing solution into a reference platform for several other programming systems, such as Storm[2] for streaming data analysis; Hive[3] for querying large datasets; Giraph[4] for iterative graph processing; HBase[5] for random and real-time read/write access to data in a non-relational model; Oozie,[6] for managing Hadoop jobs; Ambari[7] for provisioning, managing, and monitoring Hadoop clusters; and ZooKeeper[8] for maintaining configuration information, naming, and providing distributed synchronization and group services. An overview of the Hadoop software stack is shown in Figure 4.1.

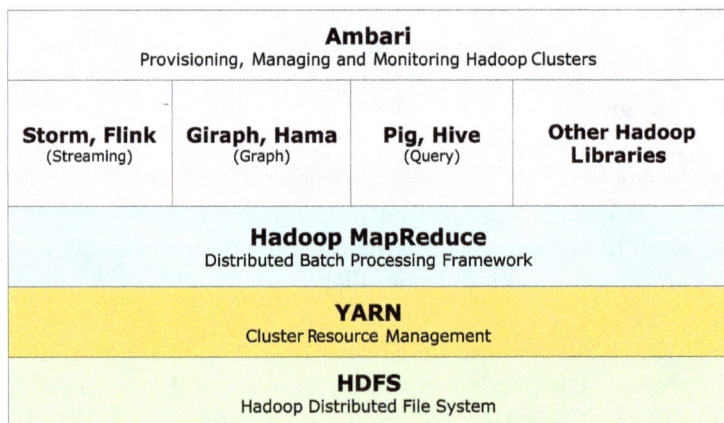

Ambari			
Provisioning, Managing and Monitoring Hadoop Clusters			
Storm, Flink (Streaming)	**Giraph, Hama** (Graph)	**Pig, Hive** (Query)	**Other Hadoop Libraries**
Hadoop MapReduce Distributed Batch Processing Framework			
YARN Cluster Resource Management			
HDFS Hadoop Distributed File System			

Fig. 4.1.   The Hadoop software stack.

[2]https://storm.apache.org/.
[3]https://hive.apache.org/.
[4]https://giraph.apache.org/.
[5]https://hbase.apache.org/.
[6]https://oozie.apache.org/.
[7]https://ambari.apache.org/.
[8]https://zookeeper.apache.org/.

4.2.1.1 *Hadoop Distributed File System*

The HDFS was designed to store large volumes of data while allowing for fast reading and fault tolerance. To this end, files in HDFS are distributed and replicated on different storage nodes, which facilitate the execution of parallel and distributed applications. It supports a hierarchical file organization similar to that of traditional file systems. It allows us to create, rename, and delete files, move a file from one directory to another, and so on. As shown in Figure 4.2, the architecture of an HDFS cluster is based on a master–worker paradigm and includes two types of nodes: a *namenode* (master) and several *datanodes* (workers).

The namenode takes care of managing the distributed file system, maintaining the file system tree and storing names and metadata of both files and directories. Conversely, the datanodes store and retrieve the data blocks, periodically communicating to the namenode the list of data blocks they are storing. Although datanodes store data, it is worth noting that the namenode is a fundamental component of the system, without which the file system cannot be used. To make the namenode fault-tolerant, the architecture also includes a *secondary namenode*, which does not represent a replica or a backup of the namenode but rather only stores the state of the file system in case of namenode errors.

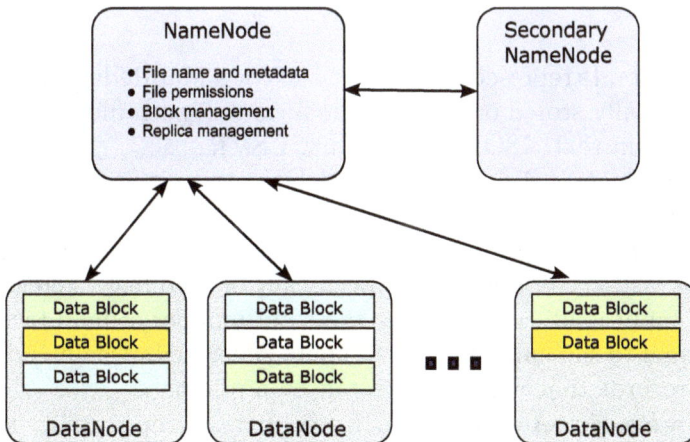

Fig. 4.2. The HDFS architecture.

HDFS is designed to store files as a sequence of data blocks, which represent the minimum amount of data that can be read or written. The default block size is 128 MB but can be configured per file. Furthermore, to ensure fault tolerance, each block is replicated among datanodes with a certain replication factor, which can be set at the time of file creation and subsequently modified. To minimize the cost of seeks, HDFS blocks are larger than blocks used in traditional file systems. For efficiency reasons, Hadoop leverages the concept of data locality to distribute compute jobs among the workers. The basic principle is that computation is much more efficient if it is performed close to the processed data. In this manner, processing does not require transferring large amounts of data over the network, minimizing network congestion and increasing overall system throughput.

4.2.1.2   *The Hadoop execution flow*

In the MapReduce programming model, data are processed in parallel by dividing the overall work into a set of independent tasks. Although the MapReduce paradigm consists of two main phases, map and reduce, the data processing flow in Hadoop is composed of different phases that use other components in addition to mapper and reducer, which can be configured by programmers. In particular, programming a Hadoop application requires specifying the following components, which are used in the execution flow as shown in Figure 4.3:

- *Input data*: It represents the input files for a MapReduce job, which are typically stored on HDFS. The format of such files is arbitrary (e.g., plain text, JSON, ZIP, binary, CSV).
- *InputFormat*: It defines how input data are split and read so as to create input splits.
- *InputSplit*: It is the portion of data that will be processed by a single instance of a mapper. Specifically, a map task is created for each input split, which means the total number of map tasks will be equal to the number of input splits generated. A split is divided into records that will be processed by a mapper.
- *RecordReader*: According to the configured input split, it converts an input split into key–value pairs suitable to be read and processed by the mapper. As an example, using the default input

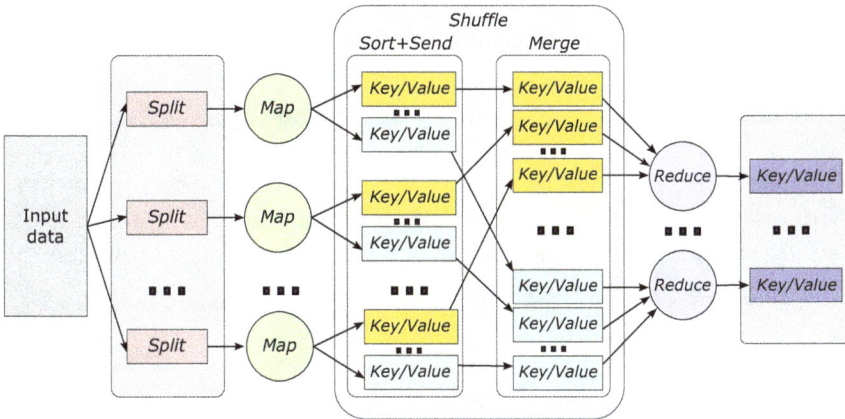

Fig. 4.3.   The Hadoop execution flow.

format (*TextInputFormat*), it creates pairs in which the key is the line number in the input file (assigned using the byte offset) and the value is the line text.

- *Mapper*: It implements the first data processing phase by applying to each key–value pair a mapping function that produces a list of key–value pairs as output.
- *Combiner*: It is a component, often referred to as *mini-reducer*, that performs local aggregation of the mapper's output. Specifically, it helps minimize the transfer of intermediate data between mappers and reducers. The output of the combiner is then passed to the partitioner.
- *Partitioner*: It takes the output from the combiner and performs partitioning by using a hashing function. In particular, incoming data are partitioned by key, so that tuples with the same key go into the same partition. Then, each partition is sent to a single reducer.
- *Shuffle and sorting*: Each partition generated by the partitioner is transferred over the network to the reduce nodes (shuffling phase). However, before sending data generated by the mapper to the assigned reducer nodes, such data are sorted by key (a process performed by the Hadoop MapReduce framework). The reducer then just merges the received sorted segments. It is worth noting that sorting in Hadoop makes it easier for the reducer to determine when to start a new task. This approach allows for time savings,

as the reducer starts a new task when the next key in the sorted data differs from the previous one.

- *Reducer*: It is the second phase of data processing in which the final aggregation is performed by applying a reduction function on data.
- *RecordWriter*: It writes the output key–value pairs from the reduce phase to output files. In particular, another component, called *OutputFormat*, defines how the output key–value pairs are written in the output files by the record writer.

4.2.1.3   *Basics*

A typical MapReduce program consists of at least three parts:

- a *mapper*, which extends the Mapper class to provide a custom implementation of the map method;
- a *reducer*, which extends the Reducer class to provide a custom implementation of the reduce method;
- a *driver*, which defines the MapReduce job and contains the main part of the program.

The *Mapper* class is defined as follows:

```
class Mapper<KEYIN, VALUEIN, KEYOUT, VALUEOUT>
```

where KEYIN, VALUEIN, KEYOUT, and VALUEOUT are the types of the input key, input value, output key, and output value, respectively. The class includes the following methods that can be overridden:

- void setup(Context context), which is called once at the beginning of the task.
- void map(KEYIN key, VALUEIN value, Context context), which is the map method called once for each key–value pair in the input split.
- void cleanup(), which is called at the end of the map task.

The *Reducer* class, defined as class Reducer<KEYIN, VALUEIN, KEYOUT, VALUEOUT>, includes the following methods:

- void setup(Context context).
- void reduce(KEYIN key, Iterable<VALUEIN> values, Context context), which is called for each key to process all the values associated with that key.
- void cleanup(), which is called at the end of the reduce task.

It is worth noting that for defining a custom combiner, it is necessary to create a reducer class. In fact, as mentioned in the previous section, the combiner is nothing more than a reducer that operates on the mapper node for efficiency reasons, as it can perform local aggregation of data produced by mappers before being sent through the network to reducer nodes.

The *Context* object allows the mapper/reducer to interact with the underlying runtime system. It includes configuration data for the MapReduce job as well as interfaces that allow it to emit output.

The *Driver* class is responsible for configuring the MapReduce job to run in Hadoop. In this class, programmers can define the job name, the types of input/output data, mapper and reducer classes, and other parameters.

Secondary sort: As default behavior, Hadoop sorts the incoming key–value tuples by key before giving them to the reducer. In some cases, however, the programmer may need to control the data ordering, such as setting them to be sorted by value rather than by key. Secondary sort is a technique that allows the MapReduce programmer to control the sorting of tuples sent to a reduce method call. It is based on the use of a composite key `<primary_key, secondary_key>`. By defining a custom partitioner, tuples with the same primary key are sent to the same reducer node. Specifically, the partitioner assigns all tuples with the same primary key to a single reducer node; then, a custom sort comparator is used to sort the tuples by both primary and secondary keys; finally, by defining a custom group comparator, tuples are grouped by primary key before being sent to the reduce method call. This sorting pattern is useful in many application cases. As an example, let us suppose we have to extract the daily routes of users from a large set of geolocalized data. Using a composite key `<user_id, timestamp>`, the data can be partitioned and grouped by user ID and sorted by timestamp (in ascending or descending order, as defined in the sort comparator class).

4.2.1.4 *Programming example*

We provide an application example that shows how Hadoop MapReduce can be exploited for creating an inverted index for a large set of web documents (Sarkar *et al.*, 2015). An inverted index contains a

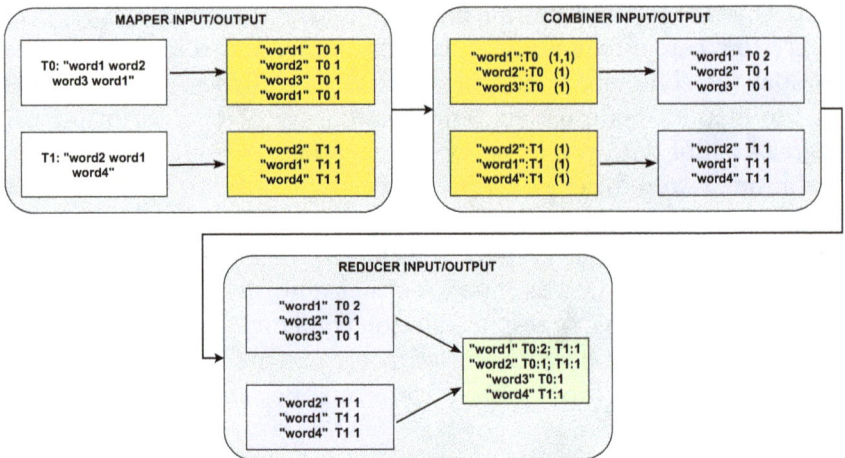

Fig. 4.4.   Execution flow of the proposed Hadoop application.

set of words (index terms), and for each word, it specifies the IDs of all the documents that contain it and the number of occurrences in each document. The inverted index data structure is a central component of a search engine indexing system. Figure 4.4 shows the data flow and main components (Mapper, Combiner, and Reducer) of the proposed application.

The MapTask class, implementing the mapper, parses the text lines coming from some input documents and emits a pair ⟨*word, documentID:numberOfOccurrences*⟩ for each word they contain, where *documentID* is the identifier of the document and *numberOfOccurrences* is set to 1 (see Listing 4.1). Each word is processed with common steps of natural language processing, such as punctuation removal, lemmatization, and stemming. To achieve a more lightweight handling of object serialization, programmers need to use specific types for keys and values. As an example, Hadoop uses *Text* and *IntWritable* instead of String and Integer, respectively, which contain the same information by using a much easier abstraction on top of byte arrays.

```
public class MapTask extends Mapper<Object, Text, Text, Text> {
    private final Text keyContent = new Text();
    private final Text valueContent = new Text();
    private final static IntWritable one = new IntWritable(1);
```

```
@Override
protected void map(Object key, Text value, Context context)
    throws IOException, InterruptedException {
    // Extract the filename from the current input split
    FileSplit fileSplit = (FileSplit) context.getInputSplit();
    String filename = fileSplit.getPath().getName();
    StringTokenizer it = new StringTokenizer(value.toString());
    while (it.hasMoreTokens()) {
        // Remove punctuation, apply lemmatization and stemming
        String word = process(it.nextToken());
        keyContent.set(word);
        valueContent.set(filename + ":" + one);
        context.write(keyContent, valueContent);
    }
}
}
}
```

Listing 4.1: Inverted index Mapper.

After word mapping, a combiner is used to aggregate intermediate data produced by the mappers before passing them to reducers. As shown in Listing 4.2, the CombineTask class implements the combiner logic by summing all the occurrences of each word that appear multiple times in a document and emits a list of pairs ⟨*word, documentID*:*sumNumberOfOccurrences*⟩.

```
public class CombineTask extends Reducer<Text, Text, Text, Text> {
    private final Text sumContent = new Text();

    @Override
    protected void reduce(Text key, Iterable<Text> values,
        Context context) throws IOException, InterruptedException {
        // Sum all the occurrences of a word in the document
        HashMap<String, Integer> sumMap = new HashMap<>();
        for (Text value : values) {
            String[] parts = value.toString().split(":");
            sumMap.merge(parts[0], 1, Integer::sum);
        }
        for (Map.Entry<String, Integer> e : sumMap.entrySet()) {
            sumContent.set(e.getKey() + ":" + e.getValue());
            context.write(key, sumContent);
        }
    }
}
}
```

Listing 4.2: Inverted index Combiner.

For each word, the ReducerTask, implementing the reducer, produces the list of all the documents containing that word and the number of occurrences in each document. Specifically, as shown in Listing 4.3, a ⟨*word, List(documentID:numberOfOccurrences)*⟩ pair is emitted for each word. The set of all output pairs generated by the reduce function forms the inverted index for the input documents.

```
public class ReduceTask extends Reducer<Text, Text, Text, Text> {
    private final Text result = new Text();

    @Override
    protected void reduce(Text key, Iterable<Text> values, Context
        context) throws IOException, InterruptedException {
        StringBuilder fileList = new StringBuilder();
        HashMap<String, Integer> sumMap = new HashMap<>();
        for (Text value : values) {
            String[] parts = value.toString().split(":");
            sumMap.merge(parts[0], Integer.parseInt(parts[1]),
                Integer::sum);
        }
        for (Map.Entry<String, Integer> e : sumMap.entrySet()) {
            fileList.append(e.getKey() + ":" + e.getValue())
                .append(";");
        }
        result.set(fileList.toString());
        context.write(key, result);
    }
}
```

Listing 4.3: Inverted index Reducer.

Finally, Listing 4.4 shows the main class (i.e., the driver) used to set up and run the application. A programmer must specify the classes to be used as mapper, combiner, and reducer, the input/output format of such classes, and the data input/output paths.

```
public class InvertedIndexJob extends Configured implements
    Tool {

    @Override
    public int run(String[] args) throws Exception {
        Configuration conf = new Configuration();
        Job job = Job.getInstance(conf);
        job.setJarByClass(this.getClass());
        job.setMapperClass(MapTask.class);
        job.setCombinerClass(CombineTask.class);
```

```
job.setReducerClass(ReduceTask.class);
// Set the output class of key and value for the
    mapper
job.setMapOutputKeyClass(Text.class);
job.setMapOutputValueClass(Text.class);
// Set the output class of key and value for the
    reducer
job.setOutputKeyClass(Text.class);
job.setOutputValueClass(Text.class);
// Specify input and output paths
FileInputFormat.addInputPaths(job,"webPage1,
    webPage2,...");
FileOutputFormat.setOutputPath(job,
    new Path(args[0]));
System.exit(job.waitForCompletion(true) ? 0 : 1);
    }
}
```

Listing 4.4: Inverted index job.

4.3 Workflow-Based Programming Tools

This section discusses some frameworks that support the workflow programming model. Due to their ability to model diverse and complex scenarios, workflows are used in a wide range of application domains, including scientific simulation, data analytics, and machine learning. For this reason, various parallel and distributed frameworks have been proposed that exploit this programming model and, in particular, the directed acyclic graph (DAG) abstraction to model execution. Among the workflow-based frameworks, *Apache Spark*[9] is one of the most popular open-source solutions, especially for developing general-purpose applications. Other popular frameworks that take advantage of the workflow paradigm in the field of streaming data analysis are *Apache Storm*[10] and *Apache Flink*.[11]

Some effort has also been devoted to the development of frameworks to facilitate the design and execution of workflow-based applications in order to efficiently exploit distributed computational and/or storage resources. Among these solutions, *Apache Airflow*,[12]

---

[9]https://spark.apache.org/.
[10]https://storm.apache.org/.
[11]https://flink.apache.org/.
[12]https://airflow.apache.org/.

COMPSs (Lordan *et al.*, 2014), the *Data Mining Cloud Framework (DMCF)* (Marozzo *et al.*, 2018), *Kepler* (Ludäscher *et al.*, 2006), *YAWL* (Yet Another Workflow Language) (Van Der Aalst and Ter Hofstede, 2005), and *Pegasus* (Deelman *et al.*, 2005) are especially used to facilitate the development and execution of scientific data analysis workflows in distributed environments.

In the following, we discuss in detail three of these frameworks as representative of workflow-based programming tools: Apache Spark, Apache Storm, and Apache Airflow.

4.3.1   *Apache Spark*

Apache Spark is a unified analytics engine designed for large-scale data analysis (Salloum *et al.*, 2016). Initially developed in 2009 by Matei Zaharia at UC Berkeley's AMPLab, in 2013, Spark joined the Apache Software Foundation. With its in-memory programming feature and higher-level modules for several workloads that can be combined in the same application, Spark has established itself as the *de facto* framework for big data analytics. In fact, many powerful and robust libraries are built on top of Spark, making it a flexible system for a wide range of applications, such as Spark SQL[13] for dealing with SQL queries, MLlib[14] for scalable machine learning applications, GraphX[15] for graph-parallel computation, and Spark Streaming[16] for streaming analysis. The execution of a generic Spark application on a cluster is driven by a central coordinator (i.e., the main process of the application), which can connect with different cluster managers, such as Apache Mesos,[17] YARN, or Spark Standalone (i.e., a cluster manager embedded into the Spark distribution). For provisioning, managing, and monitoring Spark clusters, Ambari serves as a comprehensive solution. Although Spark does not provide its own distributed storage system, it has been designed to run on top of several data sources, such as distributed file systems (e.g., HDFS), cloud object storages (e.g., Amazon S3, OpenStack Swift),

---

[13]http://spark.apache.org/sql/.
[14]https://spark.apache.org/mllib/.
[15]http://spark.apache.org/graphx/.
[16]https://spark.apache.org/streaming/.
[17]http://mesos.apache.org/.

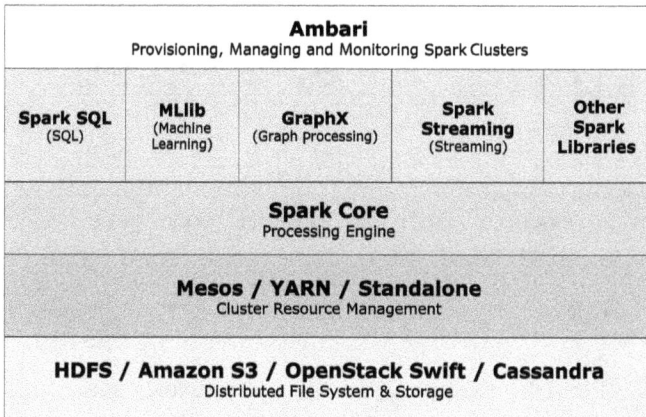

Fig. 4.5.   The Spark software stack.

and NoSQL databases (e.g., Cassandra). A comprehensive software stack is shown in Figure 4.5.

Several big companies, including eBay, Amazon, and Alibaba, use Spark in production to quickly extract insights from data for analysis purposes. For example, eBay uses big data and machine learning solutions based on Spark for log aggregation and providing targeted offers to enhance customer experience. Thanks to a very large community of users and contributors, the development of Spark is constantly expanding. In particular, many efforts are oriented toward the MLlib library, which provides advanced data analytics with parallel machine learning algorithms.

4.3.1.1   *Main concepts*

The Spark core provides basic functionalities, such as task scheduling, memory management, fault recovery mechanisms, and different abstractions for representing data and controlling computation. In particular, data are represented as resilient distributed datasets (RDDs), whereas computations on these RDDs are expressed as transformations or actions.

Resilient distributed datasets: An RDD is a distributed memory abstraction that represents a collection of items distributed across the computing nodes of a cluster that can be manipulated in parallel (see Figure 4.6). These items are immutable, fault-tolerant, and

Logical view

| 1 | 2 | ▪ ▪ ▪ | 1000 | 1001 | 1002 | ▪ ▪ ▪ | 2000 | ▪ ▪ ▪ | 9999 | 10000 |

Physical view

| 1 | 2 | ▪ ▪ ▪ | 1000 | | 1001 | 1002 | ▪ ▪ ▪ | 2000 | ▪ ▪ ▪ | | ▪ ▪ ▪ | 9999 | 10000 |

Partition1   Partition2   ▪ ▪ ▪   Partition10

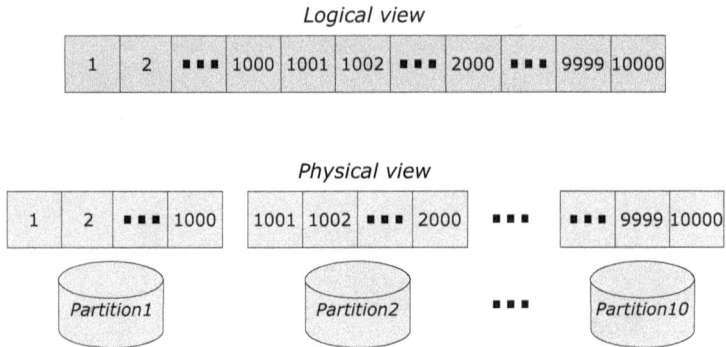

Fig. 4.6.   Example of a resilient distributed dataset.

partitioned in such a way that at least one partition is stored in the main memory of each cluster node, or if there is not enough memory, it is stored on the local disk through a spilling operation. A partition is an atomic chunk of data, and each worker node can run the code specified by the application on the partition of the RDD it owns. Once created, the content of an RDD cannot be modified, but any modification creates a new RDD. This also allows for automatic recovery in case of failure, using lineage information to recompute lost data in an efficient way. This operation is completely hidden from the user because for each RDD, Spark knows how it has been created and in case of failure, RDDs are automatically rebuilt using the RDD lineage DAG.

Transformations and actions: A Spark application is written as a sequence of operations to be performed on RDDs. In particular, RDDs can be manipulated through two types of operations:

- *Transformations*, which are coarse-grained operations that create new RDDs based on previous ones. Examples are *map*, *filter*, and *join* operations.
- *Actions*, which perform a computation on an RDD or write data to the storage system. These operations generally output a value, such as *count, collect, reduce,* and *save.*

A Spark application at runtime is represented as a DAG composed of data sources, transformations and operations, and data sinks. An important aspect of transformations on RDDs is that they are evaluated in a lazy way (*lazy evaluation*), which means that a new

RDD is not immediately computed but is materialized only when an action is triggered. This allows Spark to perform optimization in the DAG of operations, e.g., merging multiple filter/map transformations or avoiding moving data for a group transformation if data are partitioned.

Transformations and actions make it easy to code a parallel application. In fact, compared to Hadoop, developing an application using Spark results in a smaller number of lines of code, especially when used with the Scala programming language, which provides an object-oriented and functional programming high-level interface. In this sense, Spark provides a *medium level of abstraction* because programmers are equipped with powerful APIs that hide low-level details related to parallel and distributed computing. However, medium programming skills are required, mainly related to functional programming.

Data locality: In most cases, code and data are separated, creating the need to move between them. Data locality refers to the process of moving computation close to where the actual data resides on the cluster's node, instead of moving large data to computation. Typically, it is faster to ship serialized code from place to place than a chunk of data because the code size is much smaller than data. This minimizes network congestion and increases the overall throughput of the system. Spark builds its scheduling around this general principle of data locality (see Figure 4.7). It turns out that reading data within the same node directly from the main memory (alternatively from disk) is faster than reading between nodes of the same rack, while it is even slower to read between nodes in different racks (data access rate). In situations where there is no unprocessed

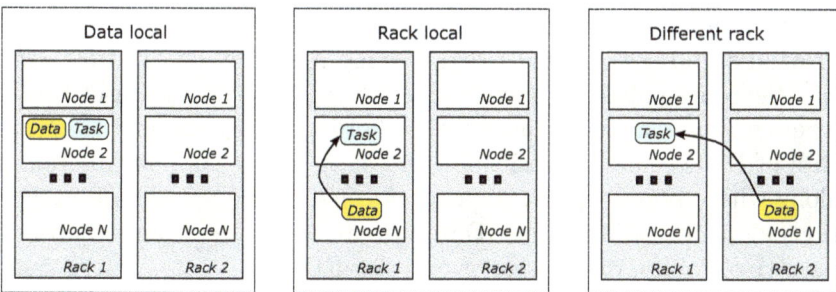

Fig. 4.7. Data locality.

data on any idle executor, a decision must be made: (a) wait until a busy CPU frees up to start a task on data on the same server, or (b) immediately start a new task in a farther away place that requires moving data there.

Caching: As already partially discussed, Spark also supports the persistence of RDDs in main memory. This operation, called caching or in-memory computing, allows for future reuse of RDDs, making Spark up to 100× faster than Hadoop (Verma *et al.*, 2016), especially for iterative algorithms. When an RDD is persisted, each worker stores in the main memory the partition it owns and reuses it for other actions. This mechanism is also fault-tolerant due to the RDD lineage, which allows for automatic recomputing of an RDD partition if it is lost. Spark tries to keep in memory as much data as possible, using efficient serialization libraries (e.g., Kryo) to reduce their memory occupation. However, if memory is saturated, the RDD may be spilled to disk. For this reason, even though Spark can be considered a better alternative to Hadoop, in some classes of applications it has its limitations that make it complementary to Hadoop. The main one is that data must fit in the main memory to reduce execution time. In fact, RAM is a critical resource and Spark can suffer from the lack of automatic optimization processes aimed at maximizing in-memory computing while minimizing the probability of data spilling, which is a major cause of performance degradation (Cantini *et al.*, 2021).

DataFrames and Datasets: Spark introduced a new data abstraction called DataFrame in version 1.3 and later another abstraction named Dataset in version 2.0. Similarly to RDDs, DataFrames and Datasets are immutable and distributed data collections. In addition, DataFrames organize data into named columns, like tables in relational databases. Since Spark 2.0, Datasets extend DataFrames and provide an object-oriented programming interface, i.e., DataFrames can be represented as a collection of generic type Dataset[Row], where a *Row* is a generic and untyped object. Datasets offer the benefits of RDDs, such as resilience and support for lambda functions, along with a set of optimizations performed by the Spark SQL execution engine. DataFrames and Datasets can be created from structured files, Hive tables, databases, or existing RDDs.

Broadcast variables and accumulators: Broadcast variables and accumulators are used to share variables across cluster nodes. Broadcast variables enable the programmer to cache a read-only variable on each node rather than delivering a copy. They can be used to efficiently distribute a copy of a big dataset to each node, using efficient broadcast algorithms. Instead, accumulators are similar to global variables that can be efficiently handled in parallel since they are only applicable to operations that are associative and commutative. For example, they can be used for aggregating information across the executors, such as counters or sums.

4.3.1.2 *Architecture*

As shown in Figure 4.8, the architecture of Apache Spark consists of a master node that runs the driver program responsible for executing the application submitted by the user and a set of worker nodes. Spark employs a master–worker architecture with the driver as the central coordinator and the worker nodes that run one or more executors responsible for running tasks assigned by the driver and storing data.

In particular, a Spark application is defined as a set of independent *stages* running on a pool of worker nodes and connected in a DAG. A stage is a set of *tasks* executing the same code on different partitions of input data, thus providing *data parallelism*, as input

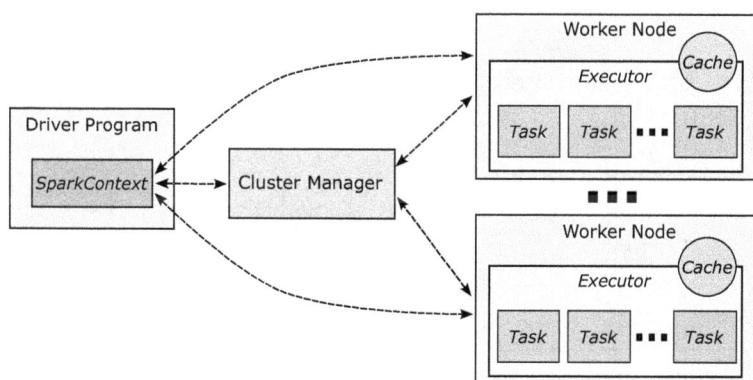

Fig. 4.8. Architecture of a Spark cluster.

data are divided into chunks and processed in parallel by different computing nodes. Spark supports *task parallelism* as well when independent stages of the same application are executed in parallel.

The driver program is also responsible for initializing the Spark context, which acts as the entry point to a Spark cluster. The Spark context can communicate with a variety of cluster managers for acquiring resources on the cluster, such as Mesos, YARN, and Kubernetes, and once connected, it submits the application (i.e., a JAR or a Python file) to the executors.

It is worth noting that each application submitted to the driver has its own pool of executor processes, which perform tasks using threads. This allows for separating applications from one another, both in terms of scheduling and execution. However, this implies that data cannot be exchanged among different applications defined with different instances of the Spark context but need to be written to an external storage system. Figure 4.9 shows how tasks are generated from a Spark application. An application creates RDDs, transforms them, and runs actions. This results in a DAG of operators. The DAG is compiled into stages, which are sequences of RDDs without shuffle operations in between. Each stage is executed as a set of *tasks*, with typically one task generated for each partition.

4.3.1.3   *Basics*

A typical Spark application requires the definition of the SparkContext, which is the entry point for Spark functionality. It represents

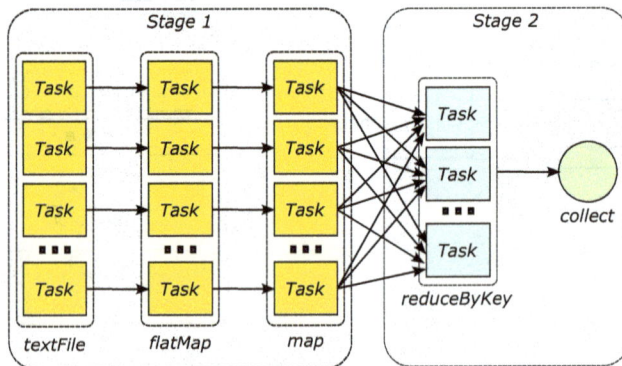

Fig. 4.9.   Spark's task flow.

the connection to a Spark cluster and can be used to create RDDs, accumulators, and broadcast variables. The Spark driver program uses the SparkContext to connect to the cluster through a resource manager (e.g., YARN or Mesos). A SparkConf object is required to create the SparkContext, storing configuration parameters such as the name of the application and the master URL (i.e., a Spark, Mesos or YARN cluster URL, or a special "local" string to run an application in local mode). Other parameters can be included in the configuration, such as the number of cores and the memory size of the executor running on a worker node.

```
val conf = new SparkConf().setAppName(appName).
  setMaster(master)
new SparkContext(conf)
```

Since version 2.0, SparkSession provides a single point of entry to interact with underlying Spark functionalities and allows programmers to use DataFrame and Dataset APIs, as well as the APIs of SQL, streaming, and graphs. All the functionalities available with SparkContext are also available in SparkSession.

```
new SparkSession.builder.appName(appName).getOrCreate()
```

Once the SparkContext or SparkSession is created, programmers can create data collections that will be parallelized in the cluster. Parallelized collections are created by invoking the *parallelize* function on an existing collection (e.g., an array), and its items will be replicated to create a distributed dataset (i.e., an RDD). Programmers can also specify the number of partitions to divide the dataset into.

```
val data = Array(1, 2, 3, 4, 5)
val distData = sc.parallelize(data)
```

Distributed datasets can also be created from any Hadoop storage source, including the local file system (the file must be accessible by worker nodes), HDFS, HBase, Amazon S3, and others. For example, a test file can be loaded into an RDD using the *textFile* method.

```
val distFile = sc.textFile("data.txt")
```

Once loaded, programmers can apply transformations and actions on such distributed data using the *functional programming* style of the Scala language. For example, the following shows how a text file, which can be loaded with the previous instruction, can be processed to calculate its length, using a *map* and a *reduce* function. It is worth noting that the result of the map transformation is not immediately computed because Spark adopts a lazy evaluation strategy. Only when the reduce action is called, Spark performs the computation by distributing tasks on all worker nodes and then aggregating partial results into the driver program.

```
val lineLengths = distFile.map(s => s.length)
val totalLength = lineLengths.reduce((a, b) => a + b)
```

Some of the most common transformations and actions supported by Spark are listed in Table 4.1. For each of them, input and output data types are described. The meaning of each operation is detailed in Table 4.2.

Table 4.1.   Common transformations and actions supported by Spark.

Transformations	map(f: T $\Rightarrow$ U) : RDD[T] $\Rightarrow$ RDD[U]
	filter(f: T $\Rightarrow$ Bool) : RDD[T] $\Rightarrow$ RDD[T]
	flatMap(f: T $\Rightarrow$ Seq[U]) : RDD[T] $\Rightarrow$ RDD[U]
	sample(fraction: Float) : RDD[T] $\Rightarrow$ RDD[T]
	groupByKey() : RDD[(K, V)] $\Rightarrow$ RDD[(K, Seq[V])]
	reduceByKey(f: (V, V) $\Rightarrow$ V) : RDD[(K, V)] $\Rightarrow$ RDD[(K, V)]
	union() : (RDD[T], RDD[T]) $\Rightarrow$ RDD[T]
	join() : (RDD[(K, V)], RDD[(K, W)]) $\Rightarrow$ RDD[(K, (V, W))]
	partitionBy(c: Partitioner[K]) : RDD[(K, V)] $\Rightarrow$ RDD[(K, V)]
Actions	count() : RDD[T] $\Rightarrow$ Long
	collect() : RDD[T] $\Rightarrow$ Seq[T]
	reduce(f: (T, T) $\Rightarrow$ T) : RDD[T] $\Rightarrow$ T
	take(n: Int) : RDD[T] $\Rightarrow$ Seq[T]

4.3.1.4   *Programming example*

The programming example discussed here is a simplified version of a market basket analysis task that looks for products that frequently appear together in a set of transactions. In particular, given a text file of transactions represented as a set of items (one for each row) delimited by a comma, the program outputs all pairs of items ordered by the number of co-occurrences (see Figure 4.10).

Table 4.2.   Meaning of common transformations and actions supported by Spark.

Transformation/Action	Meaning
map(f)	Return a new RDD computed by passing each element of the source through a function *f*.
filter(f)	Return a new RDD that contains those elements of the source on which *f* returns true.
flatMap(f)	Similar to map, but each input item can be mapped to zero or more output items.
sample(fraction)	Sample a fraction of the data, with or without replacement.
groupByKey()	Group the values for each key in the RDD into a single sequence.
reduceByKey(f)	Merge the values for each key using an associative and commutative reduce function.
union()	Merge the elements in the source dataset and the argument.
join()	Return a new RDD of pairs with all pairs of elements for each key.
partitionBy(c)	Reshuffle the data in the RDD randomly to create either more or fewer partitions.
count()	Return the number of elements in the dataset.
collect()	Return all the elements of the dataset as an array at the driver program.
reduce(f)	Aggregate the elements of the dataset using a commutative and associative function.
take()	Return an array with the first n elements of the dataset.

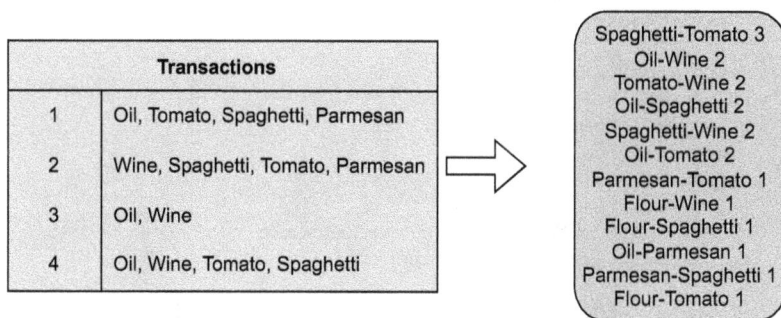

Transactions		Spaghetti-Tomato 3
1	Oil, Tomato, Spaghetti, Parmesan	Oil-Wine 2
2	Wine, Spaghetti, Tomato, Parmesan	Tomato-Wine 2
3	Oil, Wine	Oil-Spaghetti 2
4	Oil, Wine, Tomato, Spaghetti	Spaghetti-Wine 2

Spaghetti-Tomato 3
Oil-Wine 2
Tomato-Wine 2
Oil-Spaghetti 2
Spaghetti-Wine 2
Oil-Tomato 2
Parmesan-Tomato 1
Flour-Wine 1
Flour-Spaghetti 1
Oil-Parmesan 1
Parmesan-Spaghetti 1
Flour-Tomato 1

Fig. 4.10.   Ordered pairs of items that appear most together in a set of transactions.

The code is presented in Listing 4.5. As a first step, the programmer needs to initialize the SparkSession as an entry point to the Spark cluster. Then, the SparkContext of the SparkSession is used to load the input text file from the local file system.

Each line of the input file is split using the comma as a delimiter and is parsed into an array of items. Afterward, only the sets of items with at least two elements are kept using a *filter* transformation. These pairs are then passed to a custom method, named *getAllPairs*, which extracts all pairs into an Array[(String, String)] object. The result of this mapping is flattened through a *flatMap* and stored in a variable named *pairs*.

After this, similarly to the word-count application, the occurrences of each pair are calculated. First, for all pairs, a new key–value pair is generated through a *map*, where the key is a string obtained by the concatenation of the two items (i.e., *pair._1* and *pair._2*) and the value is 1. Then, the *reduceByKey* transformation merges the values for each key by applying the sum function to the values. The resulting RDD, stored in *pairsOcc*, contains unique pairs of items and their count (i.e., the number of co-occurrences).

Finally, the RDD obtained at the previous step is ordered using a *sortBy* transformation based on the number of co-occurrences. Since the *sortBy* clause applied to more than one partition may return a result that is only partially ordered, a *collect* is needed to retrieve all the elements of the RDD from all worker nodes to the driver node before printing the ordered list.

```scala
object MarketBasketAnalysis {

  def main(args: Array[String]) {
    val inputFile = "transactions.txt"
    val spark = SparkSession.builder
      .master("local[*]")
      .appName("MarketBasketAnalysis")
      .getOrCreate()
    val sc: SparkContext = spark.sparkContext
    val textFile = sc.textFile(inputFile)

    val items:  RDD[Array[String]] = textFile.map(line => line.
        split(','))

    val pairs:  RDD[(String, String)] = items.filter(items => items
        .length >= 2).flatMap(items => getAllPairs(items))
```

```
val pairsOcc:  RDD[(String, Int)] = pairs.map(pair => (pair._1
   + "-" + pair._2, 1)).reduceByKey((i, j) => i + j)

val ordPairsOcc:  RDD[(String, Int)] = pairsOcc.sortBy(x => x.
   _2, ascending = false)
ordPairsOcc.collect().foreach(x => println(x._1 + " " + x._2))
}

def getAllPairs(items: Array[String]): Array[(String, String)] =
   {
   for (i1 <- items; i2 <- items; if (i1.compareTo(i2) < 0))
      yield (i1, i2)
   }
}
```

Listing 4.5: Market basket analysis.

Since Spark provides APIs for structured data and SQL-like queries, we discuss another example that uses the DataFrame abstraction.

Given a *json* file that stores data about employees of an IT company (e.g., their personal information, department, salary, and skills), the aim is to perform some queries using both the methods provided by the DataFrame APIs and SQL statements.

As introduced in Section 4.3.1.1, DataFrames allow us to handle structured data and organize data into named columns, similar to tables in relational databases. To do this, Spark allows programmers to define the schema of the data that will be loaded in order to give a name to each column and refer to it using that name. Using the *StructType* object, as shown in Listing 4.6, a DataFrame is defined whose columns are the IDs of the employees, their names, surnames, and ages, the department they belong to, their salaries, and a set of skills in an array format (e.g., Java, Python, or C). For each column, the programmer can specify its format (i.e., integer, string, or array), which will be used to parse the input data. The schema is then used to load the input data from a *json* file.

```
object DataFrameExample {
def main(args: Array[String]): Unit = {
   // Initialize SparkSession

   val schemaEmpl = new StructType()
      .add("id", IntegerType)
```

```
    .add("name", StringType)
    .add("surname", StringType)
    .add("age", IntegerType)
    .add("department", StringType)
    .add("salary", IntegerType)
    .add("skills", ArrayType(StringType))
  val dfEmpl = spark.read.schema(schemaEmpl).json(fileEmpl)

  dfEmpl.groupBy("department").count().show()

  dfEmpl.agg(min("age"), max("age"), mean("age")).show()

  dfEmpl.filter("age >= 30 AND age < 40").agg(mean("salary")).
      show()
  dfEmpl.filter("age >= 40 AND age < 50").agg(mean("salary")).
      show()
  }
}
```

Listing 4.6: Analysis of a DataFrame.

Listing 4.6 shows how to perform some queries using the methods provided by the DataFrame APIs. For example, after loading the input data, we count the number of employees for each department using the *groupBy* method. Then, using the *agg* method, the entire dataset is aggregated without groups to compute the *min*, *max*, and *mean* ages of the employees. Finally, the mean salary of employees with ages in the range of 30–39 and 40–49 is computed.

Another way to perform queries is to use SQL statements directly, as shown in Listing 4.7. To do this, Spark requires creating a temporary view of the table for running SQL queries on top of that, using the *createOrReplaceTempView*. In this way, the table can be used to retrieve data in SQL statements, such as selecting all the employees of the company or, as also done in Listing 4.6, counting the number of employees for each department using the *groupBy* clause.

```
  dfEmpl.createOrReplaceTempView("itcompany")
  spark.sql("SELECT * FROM itcompany").show()
  spark.sql("SELECT department, COUNT(*) FROM itcompany
      GROUP BY department").show()
```

Listing 4.7: Analysis of a DataFrame using SQL statements.

Now, suppose we have another dataset with data about projects the IT company is involved in. For each project, the following information is described: the project ID, its name, description, budget, skills required for that project, and employee IDs, specified as an array. To know which departments of the company are most involved in a given project (e.g., the project with ID equal to 0), it is necessary to perform a *join* between the DataFrame of the employees, which contains the information on the departments, and the DataFrame of the projects with the information on the employees involved in each project. Before the two DataFrames are joined, an *explode* operation is required to return a new row for each element in the given array of employee IDs. This way, we can merge the two DataFrames with the condition that the ID of the employee in the first DataFrame is equal to that of the employee involved in a project from the second DataFrame, and we can perform a *groupBy* based on the department. The code is shown in Listing 4.8.

```
val dfProjExpl = dfProj.filter(dfProj("id") === "0").
    withColumn("employee", explode($"employees"))
dfEmpl.join(dfProjExpl, dfEmpl("id") === dfProjExpl
    ("employee")).groupBy("department").count().show()
```

Listing 4.8: Join between two DataFrames.

4.3.2 *Apache Storm*

Apache Storm is a distributed real-time computation system that allows for the processing of unbounded streams of data in a reliable way. It was originally created by Nathan Marz in 2010, with the idea of a stream processing system that could be developed in a single program as a single abstraction, and joined the Apache Software Foundation in 2014.

Before Storm, real-time processing systems were developed using queues for writing data and workers to read and process those data. Workers could send messages to other workers through queues for further processing. This approach required making sure that the queues and workers were always alive, making it difficult to build applications. Most of the logic of the application had to do with where to send/receive messages and how to serialize/deserialize messages, and not with the business. When Storm was proposed, it proved to be extremely scalable, easy to use, and capable of processing data with low latency. Nowadays, Storm is widely used for real-time analytics by big companies, such as Twitter, Groupon, and Spotify. For example, Twitter developers use Storm for processing many terabytes of data flows a day, for filtering and aggregating contents, or for applying machine learning algorithms on data streams. Its user community is relatively small; however, thanks to its user-friendly features and flexibility, Storm can be adopted by medium-sized companies for business purposes (e.g., real-time customer services, security analytics, and threat detection). Other typical use cases of Storm are online machine learning, continuous computation, and distributed RPC.

Storm provides a *medium level of abstraction*, as programmers can easily define an application by using basic abstractions (i.e., spouts, streams, bolts, and topologies, which are discussed in the following) and test it in local mode without having to run it on a cluster. Storm is written mainly in Clojure, a dialect of Lisp, but to support a very large number of potential users it provides APIs also in Java. Moreover, different programming languages are supported through the multi-language protocol that allows the implementation of bolts and spouts with other languages, including non-JVM-based languages, such as Python. The runtime of Storm supports *data parallelism* when many threads execute in parallel the same code on different chunks and *task parallelism* when different spouts and bolts run in parallel.

4.3.2.1 *Main concepts*

Data and computation abstractions: The programming paradigm offered by Storm is based on five abstractions for data and computation:

- *Tuple*: It is the basic unit of data that can be processed by a Storm application. A tuple consists of a list of fields (e.g., byte, char, integer, long).
- *Stream*: It represents an unbounded sequence of tuples, which is created or processed in parallel. Streams can be created using standard serializers (e.g., integers, doubles) or with custom ones.
- *Spout*: It is the data source of a stream. Data are read from different external sources, such as social network APIs, sensor

networks, and queuing systems (e.g., Java Message Service, Kafka[18], Redis[19]). Then, they are fed into the application.

- *Bolt*: It represents the processing entity. Specifically, it can execute any type of task or algorithm (e.g., data cleaning, functions, joins, queries).
- *Topology*: It represents a job. A generic topology is configured as a DAG, where spouts and bolts represent the graph vertices and streams act as their edges. It may run forever until it is stopped.

Storm's basic principle is that streams are created and processed in parallel, but they can be represented in a single program as a single abstraction. That inspired the concept of spouts and bolts: a spout creates new streams, whereas a bolt accepts streams as input and generates streams as output. The essential idea was that spouts and bolts were inherently parallel, just like mappers and reducers in Hadoop. Bolts simply subscribe to any streams they require and specify how the incoming stream should be partitioned. At a higher level of abstraction, the topology is a network of spouts and bolts. Part of creating a topology is defining which streams each bolt should receive as input via a stream grouping that determines how the stream should be partitioned across bolts. Shuffle, fields, and direct grouping are examples of naive stream grouping:

- In *shuffle grouping*, tuples are randomly divided across bolts such that each bolt receives an equal amount of tuples.
- In *field grouping*, tuples are partitioned based on the field supplied in the grouping (i.e., tuples with the same field are assigned to the same task, whereas tuples with different fields may be processed by different tasks).
- Finally, in *direct grouping*, the producer of the tuple determines which task of the consumer will get this tuple.

Other built-in stream groupings in Storm are partial key, all, global, none, and local grouping.

---

[18]https://kafka.apache.org/.
[19]https://redis.io/.

Message processing: The strength of Storm processing lies in its algorithm designed to ensure message processing reliability. Typically, guaranteeing message processing involves intermediate brokers capable of recovering messages if a processing unit fails. However, introducing these brokers in a real-time system would complicate the architecture, making fault tolerance harder and slowing down message transit between different components (i.e., spouts and bolts). To ensure message processing and reliability without intermediate brokers, retries need to originate from the source (i.e., the spouts). However, failures can occur at any point in the system, and it's crucial to accurately identify these failures. To address these challenges, Storm uses a series of tasks to track the DAG of tuples generated by spouts. In particular, a tracker task acknowledges the spout that produced the tuple. Moreover, to deal with large tuple graphs, Storm employs a tracking algorithm that requires minimal space (around 20 bytes) for each spout tuple. Each tracker uses a map that connects a spout tuple ID to a pair of values, i.e., the task ID generating the spout tuple, which is used for acknowledgment, and a 64-bit integer, called the "ack val", which represents the overall state of the tuple tree. Particularly, this value is derived from the XOR operation of all tuple IDs in the tree that have been generated and/or acknowledged. In this way, when an acker task checks that an "ack val" has become zero, it knows that the tuple tree is completed. This mechanisms, known as "at least once" processing, ensures a stateless processing semantic by guaranteeing that all messages will be processed, though some may be processed multiple times in case of system failures. For programmers requiring stateful operations, Storm offers Trident, a micro-batching API built on top of Storm, providing an "exactly once" processing semantic.

Isolation scheduler for multi-tenant processing: Another important aspect of Storm is its multi-tenant approach for processing, i.e., the ability to run independent applications on a shared cluster, ensuring that all applications have enough resources without being affected by other applications on the cluster. In Storm, this is done by an isolation scheduler that makes it simple and secure to share a cluster across several topologies. The isolation scheduler allows the user to define which topologies should operate on a separate group of machines within the cluster where no other topologies will run (i.e., be isolated). These isolated topologies have priority on the cluster; therefore, if there is competition with non-isolated topologies, resources will be assigned to isolated ones. Once all isolated topologies have been assigned, the remaining machines in the cluster are shared among all non-isolated topologies. The isolation scheduler solves the multi-tenancy problem by providing full isolation between topologies and preventing resource conflict between them.

4.3.2.2 *Architecture*

Storm can be considered for many reasons the counterpart of Hadoop for real-time processing. Indeed, Storm cluster is similar to a Hadoop cluster, but whereas on Hadoop there are MapReduce jobs, on Storm there are topologies, with the difference that a MapReduce job eventually finishes, while a topology processes messages forever or until it is killed. A Storm cluster is composed of two types of nodes: the master node and the worker nodes. The former runs a daemon called Nimbus, which is responsible for distributing code around the cluster, assigning tasks to machines, monitoring tasks for failures, and restarting them if required. Conversely, each worker node runs a daemon called Supervisor, which listens for jobs assigned to its machine and starts and stops worker processes as necessary based on what Nimbus has assigned to it. Each worker process executes a subset of a topology, which consists of many worker processes across many machines. Figure 4.11 shows the architecture of a typical Storm cluster.

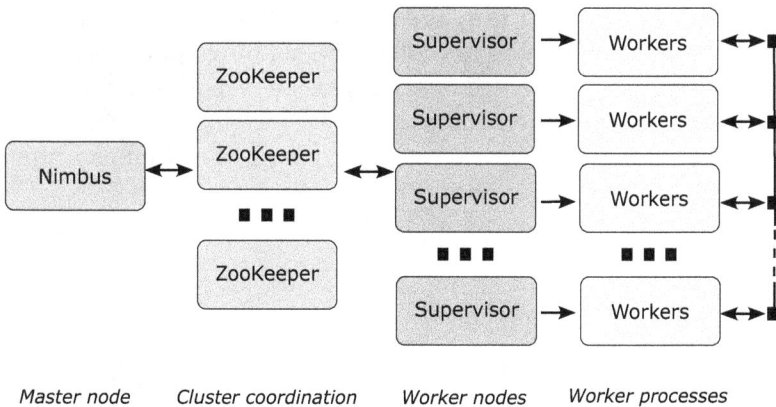

Fig. 4.11. Architecture of a Storm cluster.

Storm is designed to be a fault-tolerant system. If a worker dies, the Supervisor simply restarts it, while if it cannot be restarted (i.e., heartbeats to Nimbus fail), the worker will be reassigned to another machine by Nimbus. Also, the Nimbus and Supervisor daemons are designed to be fail-fast and stateless, i.e., all states are kept in a component called ZooKeeper. As a consequence, if the Nimbus or Supervisor daemons die, they simply restart, and workers are not affected by their failure. In this way, there is a soft single point of failure at the Nimbus node, in the sense that without the Nimbus, the workers will still continue to work and Supervisors will continue to restart them if required. However, in the absence of Nimbus, workers will not be automatically reassigned to alternative machines when necessary. The stateless property of Nimbus and Supervisors is guaranteed by the ZooKeeper, which coordinates, shares configuration information among applications with robust synchronization techniques, and stores all of the states associated with the cluster and tasks.

To sum up, a Storm cluster should have one Nimbus node, multiple Supervisors, and one ZooKeeper instance per machine that are used to coordinate the Nimbus and the Supervisors. The entire workflow of a Storm application follows these steps:

(1) The Nimbus waits for the submission of the topology.
(2) When submitted, the Nimbus will gather all tasks of the topology and their execution order and distributes the tasks among all running Supervisors.
(3) All Supervisors send heartbeats to the Nimbus at regular intervals to confirm that they are still alive. Any failure (i.e., Supervisors or Nimbus) will not affect the running application.
(4) When all tasks have been performed, the supervisor will wait for a new task from the Nimbus.
(5) When all topologies have been processed, the Nimbus waits for a new topology to be submitted by the user.

4.3.2.3 *Basics*

A typical Storm program consists of three components:

- *spout*, which implements the *IRichSpout* interface to provide a custom implementation of the *nextTuple()* method;

- *bolt*, which implements the *IRichBolt* interface to provide a custom implementation of the *execute()* method;
- *main*, which defines the topology and the nodes of the DAG using the *setSpout* and *setBolt* methods.

The *Spout* class includes the following methods that can be overridden:

- `void open()`, which is called when a task is initialized within a worker on the cluster.
- `void nextTuple()`, which is used for emitting tuples into the topology through a collector. This method should be non-blocking, so if the Spout has no tuples to emit, it should return.
- `void ack()` and `void fail()`, which are called when Storm detects that a tuple emitted from the spout either completes successfully or fails to be completed.
- `void declareOutputFields()`, which declares the output schema for all the streams of the topology.

Instead, the *Bolt* class includes the following methods:

- `void prepare()`, which provides the bolt with an output collector that is used for emitting tuples.
- `void execute()`, which receives a tuple from one of the bolt's inputs and applies the processing logic to it.
- `void cleanup()`, which is called when a bolt is shutdown and should cleanup any resources that were opened.
- `void declareOutputFields()`, which declares the output schema for all the streams of the topology.

The *collector* object, used by both spouts and bolts, exposes the API for emitting tuples into the topology. The main difference between the collector for the *IRichSpout* interface, namely *SpoutOutputCollector*, and *OutputCollector* for *IRichBolt*, is that spouts can tag messages with IDs so that they can be acknowledged or failed.

4.3.2.4 *Programming example*

The proposed programming example shows how Storm can be used to analyze a stream of tweets for extracting the number of occurrences of each hashtag. The proposed topology, composed of one spout and two bolts, is given as follows:

- *TweetSpout* is the only data source. This spout connects to Twitter using its APIs and emits into the topology a stream of tweets generated at different timestamps.
- *SplitHashtag* receives the tuples from the spout and performs pre-processing. Specifically, it extracts the hashtags from each tweet and emits them into the topology for processing by the subsequent bolt.
- *HashtagCounter* performs the counting operation using an internal map as data structure and prints to the console the current number of occurrences of each hashtag encountered in the stream of tuples.

Twitter offers its APIs[20] as a web service-based utility for retrieving tweets published by users in real time. These APIs are available in any programming language, and among them, *twitter4j*[21] is an unofficial open-source Java library that provides a Java-based module for conveniently accessing the Twitter Streaming API through a listener-based architecture. To use the Twitter Streaming API, a Twitter developer account is needed for obtaining the *OAuth* authentication information, i.e., the *ConsumerKey*, the *CustomerSecret*, the *AccessToken* and the *AccessTokenSecret*. The *TweetSpout*, shown in Listing 4.9, uses the Twitter APIs for retrieving tweets in real time. The class inherits the *BaseRichSpout* class, which implements the discussed *IRichSpout* interface and implements the *open, nextTuple,* and *declareOutputFields* methods. In particular, in the *open* method, the collector and a queue for message queuing are initialized, and all Twitter APIs' information is set. The queue is anchored to the *onStatus* method provided by Twitter APIs so as to poll new tweets in the *nextTuple* method, which emits them into the topology.

```
public class TweetSpout extends BaseRichSpout {
    public static final String consumerKey = "...";
    public static final String consumerSecret = "...";
    public static final String accessToken = "...";
    public static final String accessTokenSecret = "...";

    private SpoutOutputCollector collector;
    private TwitterStream ts;
    private LinkedBlockingQueue queue;
```

---

[20]https://developer.twitter.com/en/docs/twitter-api.
[21]https://twitter4j.org/.

```
public void open(Map map, TopologyContext topologyContext,
    SpoutOutputCollector spoutOutputCollector) {
    queue = new LinkedBlockingQueue();
    collector = spoutOutputCollector;
    ConfigurationBuilder cb = new ConfigurationBuilder();
    cb.setOAuthConsumerKey(consumerKey)
            .setOAuthConsumerSecret(consumerSecret)
            .setOAuthAccessToken(accessToken)
            .setOAuthAccessTokenSecret(accessTokenSecret);
    ts = new TwitterStreamFactory(cb.build()).getInstance();
    ts.addListener(new StatusListener() {
        public void onStatus(Status status) {
            queue.offer(status.getText());
        }

        public void onDeletionNotice(StatusDeletionNotice
            statusDeletionNotice) {}

        public void onTrackLimitationNotice(int i) {}

        public void onScrubGeo(long l, long ll) {}

        public void onStallWarning(StallWarning stallWarn)
            {}

        public void onException(Exception e) {}
    });
    ts.sample();
}

public void nextTuple() {
    Object s = queue .poll();
    if(s == null) {
        Utils.sleep(50);
    } else {
        collector.emit(new Values(s));
    }
}

public void declareOutputFields(OutputFieldsDeclarer
    outputFieldsDeclarer) {
    outputFieldsDeclarer.declare(new Fields("tweet"));
}
```

Listing 4.9: Tweet Spout.

The *SplitHashtag* bolt, shown in Listing 4.10, extracts the hashtags from the content of the tweet and then sends each hashtag to a bolt of type *HashtagCounter*. The class inherits from the *BaseRichBolt* class, which implements the discussed *IRichBolt* interface and implements the *prepare*, *execute*, and *declareOutputFields* methods. The *execute* method is responsible for reading the input tuples (i.e., tweets), identifying the hashtags, and emitting them separately as a Tuple. The *getString* method is used to retrieve the tweet from the Tuple object. Using a Java *StringTokenizer* object, the tweet is broken into tokens, and the list of tokens of the tweet is iterated to find a hashtag, which is recognized by the first character "#". Finally, the *ack* method is called on the input tuple.

```java
public class SplitHashtag extends BaseRichBolt {
    private OutputCollector _collector;

    @Override
    public void prepare(Map conf, TopologyContext context,
        OutputCollector collector) {
        _collector = collector;
    }

    @Override
    public void execute(Tuple tuple) {
        String tweet = tuple.getString(0);
        StringTokenizer st = new StringTokenizer(tweet);
        while(st.hasMoreElements()) {
            String tmp = (String) st.nextElement();
            if(StringUtils.startsWith(tmp, "#")) {
                _collector.emit(new Values(tmp));
            }
        }
        _collector.ack(tuple);
    }

    @Override
    public void declareOutputFields
        (OutputFieldsDeclarer declarer){
        declarer.declare(new Fields("hashtag"));
        }
    }
}
```

Listing 4.10: SplitHashtag bolt.

As shown in Listing 4.11, the *HashtagCounter* maintains an instance variable of type Map⟨*String*, *Integer*⟩, which will store all the hashtag–occurrence pairs. The class inherits from *BaseRichBolt* and implements the *prepare*, *execute*, and *declareOutputFields* methods. The *prepare* method initializes the collector and the map. The *execute* method receives as input a tuple containing a hashtag, and after extracting the hashtag from the Tuple object using the *getString* method, it verifies that the hashtag is present in the map. If it is not already present, the hashtag is inserted as a key with 0 as the value, and the counter is incremented; otherwise, only the counter is incremented. Afterward, the new hashtag–occurrence pair is inserted into the map, and the input tuple is acknowledged. In this case, instead of emitting all pairs for each input tuple, the *cleanup* method is used as a convenient way to print final hashtag–occurrence pairs when the topology is shutdown, although it is generally used to release resources required by a bolt.

```java
public class HashtagCounter extends BaseRichBolt {
    Map<String, Integer> _counts;
    OutputCollector _collector;

    @Override
    public void prepare(Map conf, TopologyContext context,
        OutputCollector collector) {
        _counts = new HashMap<String, Integer>();
        _collector = collector;
    }

    @Override
    public void execute(Tuple tuple) {
        String hashtag = tuple.getString(0);
        Integer count = _counts.get(hashtag);
        if (count == null)
            count = 0;
        count++;
        _counts.put(hashtag, count);
        _collector.ack(tuple);
    }

    @Override
    public void cleanup() {
      for(Map.Entry<String, Integer> entry: _counts.
          entrySet()){
```

```
        System.out.println(entry.getKey()+" : " + entry.
           getValue());
      }
   }

   @Override
   public void declareOutputFields(OutputFieldsDeclarer
       outputFieldsDeclarer) {
       outputFieldsDeclarer.declare(new Fields("hashtag"));
   }
}
```

Listing 4.11: HashtagCounter bolt.

Finally, the *HahtagCountTopology* class, described in Listing 4.12, contains the *main* method. It is used for creating and initializing the application topology through a *TopologyBuilder* object and its methods *setBolt* and *setSpout*. Both methods define a new spout/bolt in the topology and take the following input parameters:

- a string ID, which will be used as a reference by the other components of the topology;
- an instance of the spout/bolt object;
- a *parallelismHint* value that indicates how many spouts/bolts tasks exist concurrently in the topology. If not specified, only one task for spout/bolt is created.

When setting the spouts and bolts in the *TopologyBuilder*, programmers need to specify the grouping mechanisms as well. In this example, tuples emitted from the TweetSpout to the SplitHashtag bolts can be randomly shuffled using *shuffleGrouping*, whereas tuples from the *SplitHashtag* bolts and the *HashtagCounter* bolts need to be grouped by key using a *fieldsGrouping*. This is required to ensure that the same hashtags are processed by the same bolt task. Finally, a *Config* object can be used to configure all topology parameters (e.g., number of workers, debug mode) and passed to a *StormSubmitter* to submit the topology on the Storm cluster.

```
public class HashtagCountTopology {
    public static void main(String[] args) throws Exception{

       TopologyBuilder builder = new TopologyBuilder();
```

```
builder.setSpout("spout", new TweetSpout, 3);
builder.setBolt("split", new SplitHashtag(), 3).
    shuffleGrouping("spout");
builder.setBolt("count", new HashtagCounter()).
    fieldsGrouping("split", new Fields("hashtag"));

Config conf = new Config();
String topologyName = "hashtagCounter";
StormSubmitter.submitTopology(topologyName, conf,
    builder.createTopology());
    }
}
```

Listing 4.12: Topology.

4.3.3 *Apache Airflow*

Apache Airflow[22] is an open-source platform designed to develop, schedule and monitor workflows. It can be used to create data processing applications as DAGs of tasks. The airflow scheduler executes the tasks on an array of workers, taking the dependencies specified by the DAG into account.

The Airflow project was started in October 2014 by Maxime Beauchemin at Airbnb and officially announced in June 2015. Open source since its first release, it joined the Apache Software Foundation's Incubator program in 2016 and reached the status of Top-Level project in 2019.

In Airflow, DAGs are defined as Python codes. There are numerous benefits to this approach, including the possibilities to store workflows in version control and to roll back to previous versions, to develop workflows by multiple people simultaneously, and to write tests to validate functionalities. This workflows-as-code approach serves three main Airflow principles:

- *Dynamicity*: Airflow pipelines are configured as Python codes, allowing for dynamic pipeline generation.
- *Extensibility*: Airflow contains operators to connect with many technologies, and all the components are extensible, allowing for easy adaptation to users' environments.

---

[22]https://airflow.apache.org/.

- *Flexibility*: Airflow facilitates the parameterization of workflows by utilizing the Jinja[23] templating engine.

Airflow presents a *high level of abstraction* as programmers can easily build workflows by combining a set of tasks and by specifying dependencies among them. A task can be easily defined as a custom Python function or as an instance of a predefined operator, which ensures a low level of *verbosity* by providing ready-to-use code for many common operations (e.g., *BashOperator* and *MySqlOperator* for executing bash commands and SQL queries, respectively). The runtime of Airflow supports *data parallelism*, when many tasks execute in parallel the same code on different data chunks, and *task parallelism*, when different tasks (or operators) run in parallel.

4.3.3.1 Architecture

A typical Airflow installation consists of the following components:

- *Scheduler*, which manages both triggering scheduled workflows and submitting tasks to the Executor.
- *Executor*, which handles the execution of tasks. In a default installation, this runs everything inside the Scheduler, whereas production-level Executors may push task execution out to a set of workers. A *Worker* typically refers to a process running on a node. Each worker process is responsible for executing tasks as part of the workflow execution.
- *Webserver*, which provides a *user interface* to inspect, trigger, and debug the behavior of DAGs and tasks.
- *DAG Directory*, containing the DAG files read by the Scheduler, Executor, and Workers.
- *Metadata Database*, used by the Scheduler, Executor, and Webserver to store state.

Figure 4.12 shows the interactions among the Airflow's components.

Depending on the specific setup, the Executor will also include other components to interact with its Workers, e.g., a task queue. However, from a high-level viewpoint, the Executor and its Workers can be seen as a single logical component.

---

[23]https://jinja.palletsprojects.com.

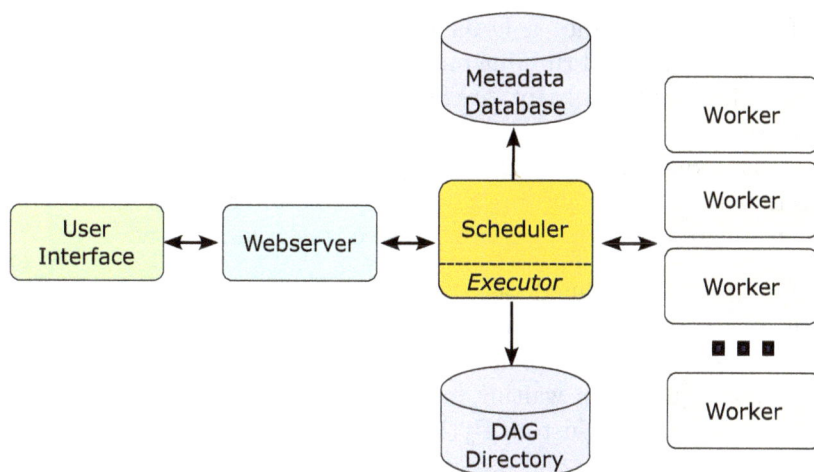

Fig. 4.12.   Architecture of an Airflow installation.

4.3.3.2   *Basics*

A DAG defines the task dependencies, their execution order, and retry behavior in Airflow. The tasks themselves describe the actions to be taken, whether it involves fetching data, running analysis, triggering other systems, or performing other operations. Listing 4.13 is a simple example of DAG declaration.

```
# DAG declaration
with DAG(dag_id="daily_backup", start_date=datetime(2023, 1, 1),
    schedule="0 0 * * *") as dag:

    # Definition of four tasks performing bash commands
    task_A = BashOperator(task_id="task_A", bash_command="mv /
        backup/*.tgz /backup/old")
    task_B = BashOperator(task_id="task_B", bash_command="tar
        czf /backup/http_log.tgz /var/log/http")
    task_C = BashOperator(task_id="task_C", bash_command="tar
        czf /backup/mail_log.tgz /var/log/mail")
    task_D = BashOperator(task_id="task_D", bash_command="echo
        backup completed")

    # Definition of task dependencies
    task_A >> [task_B, task_C]
    [task_B, task_C] >> task_D
```

Listing 4.13: Example of DAG declaration.

Listing 4.13 defines a DAG named "daily_backup," starting on January 1, 2023, and running once a day. Four tasks, all of them running a different bash script, are defined. A relationship between tasks is denoted by the symbol $>>$ to establish a dependency and determine the sequence of task execution. Airflow evaluates this script and executes the tasks according to the specified interval and established order. There are three common types of tasks in Airflow:

- *Operators*: Predefined tasks that programmers can link together quickly to build most parts of a DAG.
- *Sensors*: A special subclass of operators specifically designed for the sole purpose of waiting for an external event to occur.
- *TaskFlow* tasks: Custom Python functions packaged as tasks.

Operators: An operator is conceptually a template for a predefined task, which can simply be defined declaratively inside a DAG. Airflow offers a comprehensive set of operators available, with some built-in to the core or pre-installed providers. Some popular operators from core include:

- *BashOperator*: Executes a bash command.
- *PythonOperator*: Calls an arbitrary Python function.
- *EmailOperator*: Sends an email.

Listing 4.14 shows an example of how to use operators. In the example, an HTTP endpoint is periodically queried to get information about the current weather in a location identified by its geographical coordinates, corresponding to the town of Cosenza, Italy. This operation is carried out by `SimpleHttpOperator`, which performs a GET request toward the specified endpoint using a given query string. Then, `PythonOperator` takes the HTTP response obtained by `SimpleHttpOperator` and properly formats it to build the body of the message, which will be inserted into the email sent by `EmailOperator`.

```
endpoint = "https://api.open-meteo.com/v1/forecast"
latitude = 39.30
longitude = 16.25
parameters = ["temperature_2m_max","temperature_2m_min","
    precipitation_sum","sunrise","sunset","windspeed_10m_max","
    winddirection_10m_dominant"]
timezone = "Europe/Berlin"
today = pendulum.now().strftime("%Y-%m-%d")
```

```
weather_query = f"{endpoint}?latitude={latitude}&longitude={
    longitude}&daily={','.join(parameters)}&timezone={timezone}&
    start_date={today}&end_date={today}"

def build_body(**context):
    query_result = context['ti'].xcom_pull('submit_query')
    weather_info = json.loads(query_result)
    daily_info = weather_info["daily"].items()
    units = weather_info["daily_units"].values()
    list_info = [f"{k}:{v[0]} {unit}" for (k,v),unit in zip(
        daily_info, units)]
    body_mail = ", ".join(list_info)
    return body_mail

with DAG(dag_id="weather_mail", start_date=datetime(2023, 1, 1),
    schedule="0 0 * * *") as dag_weather:
    submit_query = SimpleHttpOperator(task_id="submit_query",
        http_conn_id='', endpoint=weather_query, method="GET",
        headers={})
    prepare_email = PythonOperator(task_id='prepare_email',
        python_callable=build_body, dag=dag_weather)
    send_email = EmailOperator(task_id="send_email", to="
        user@example.com", subject="Weather today in Cosenza",
        html_content="{{ti.xcom_pull('prepare_email')}}")

    submit_query >> prepare_email >> send_email
```

Listing 4.14: Example of operators.

It is worth noting that in this example, unlike Listing 4.13, tasks need to communicate since the *send_email* task needs the output of *prepare_email*, which in turn processes the result of the query performed by *submit_query*. Communication within Airflow relies on XCom, a mechanism that allows for the exchange of messages or small amounts of data between tasks. XCom objects are stored in the Airflow metadata database and contain an automatically generated key (which can be customized if needed), the value, and other information, such as the task and the DAG IDs. XCom internally leverages the JSON format; therefore, it requires that values must be JSON serializable. XCom objects can be explicitly pushed to (xcom_push) and pulled from (xcom_pull) the metadata database by a task instance *ti* or automatically stored as the task returns a result in output.

Sensors: Sensors are a special type of operator designed to wait for something to occur. They can be time-based, waiting for a file, or an external event. Their sole function is to wait until something happens, and upon its happening, they succeed, allowing the execution of their downstream tasks.

Because they are primarily idle, sensors have two different modes of running, allowing for increased efficiency in their utilization:

- *Poke*: The sensor takes up a worker slot for its entire runtime.
- *Reschedule*: The sensor takes up a worker slot only when it is checking and sleeps for a set duration between checks.

The poke and reschedule modes can be configured directly when the sensor is instantiated. In general, the trade-off between them is latency: an operation checking every second should be in poke mode, whereas something checking every minute should be in reschedule mode.

TaskFlow tasks: For programmers who write most of their DAGs using plain Python code rather than relying on operators, the TaskFlow API simplifies the process for developers to create well-organized DAGs using the @task decorator. Listing 4.15 is a simple example[24] showing how the Taskflow APIs can be used.

```
from airflow.decorators import task
from airflow.operators.email import EmailOperator

@task
def get_ip():
    return my_ip_service.get_main_ip()

@task
def prepare_email(external_ip):
    return {
        'subject':f'Server connected from {external_ip}',
        'body': f'Your server is connected from the external
            IP {external_ip}<br>'
    }

email_info = prepare_email(get_ip())
```

---

[24]https://airflow.apache.org/docs/apache-airflow/2.1.1/concepts/taskflow.html.

```
EmailOperator(
    task_id='send_email',
    to='example@example.com',
    subject=email_info['subject'],
    html_content=email_info['body']
)
```

Listing 4.15: Example showcasing the utilization of TaskFlow APIs.

In this example, there are three tasks: *get_ip, prepare_email,* and *send_email.* The first two are declared using TaskFlow and automatically pass the return value of get_ip into prepare_email, not just establishing the connection across XCom, but also automatically declaring that prepare_email is downstream of get_ip. send_email is a more traditional operator, yet it can still utilize the return value of prepare_email to set its parameters and again automatically determines that it must be downstream of prepare_email.

4.3.3.3 *Programming example*

In this section, we show how Airflow's TaskFlow APIs can be used to implement an ensemble learning application. Ensemble learning aims at improving classification accuracy by aggregating predictions of multiple classifiers. Among the different ways in which an ensemble model can be obtained, here we use the *voting* technique. Specifically, an ensemble method is built by fitting a set of base classifiers on training data, which performs classification on test instances by voting on the predictions made by each classifier. This classification process tends to make the ensemble classifier more accurate than any base classifier composing it. Figure 4.13 shows a schema of ensemble learning for data classification:

- The input dataset is split, using a partitioner, into a training set and a test set.
- The training set is given as input to n classification algorithms that run in parallel to build n independent classification models.
- Then, a voter tool performs an ensemble classification by assigning to each instance of the test set the class predicted by the majority of the n models generated at the previous step. For this reason, n is usually set as an odd number to avoid ties.

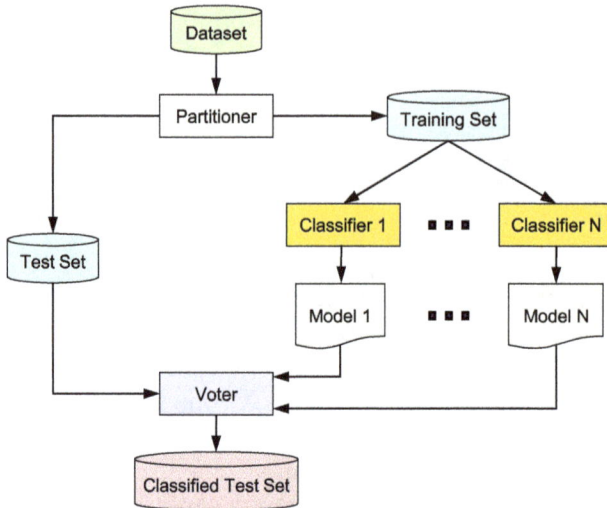

Fig. 4.13.   Ensemble learning application.

Listing 4.16 shows an implementation of the above ensemble learning application in Airflow, in which five classification models are generated.

```
# instantiate DAG
@dag(
    schedule=None,
    start_date=pendulum.now(),
    catchup=False,
    tags=["example"],
)
def ensemble_taskflow():
    # load and partition dataset into train and test
    @task(multiple_outputs=True)
    def partition():
        X,y = load_dataset(return_X_y=True)
        X_train, X_test, y_train, y_test = train_test_split(
            X, y, test_size=0.2, random_state=13
        )
        train_data = (X_train.tolist(), y_train.tolist())
        test_data = (X_test.tolist(), y_test.tolist())
        return {"train": train_data, "test": test_data}

    # fit a given classification models on train data
    @task
    def train(model:sklearn.base.BaseEstimator, train_data:tuple):
```

```
        X_train, y_train = train_data
        # fit the model and serialize it
        model.fit(X_train, y_train)
        model_bytes = pickle.dumps(model)
        model_str = model_bytes.decode("latin1")
        return model_str

    # perform ensemble classification on test data
    @task
    def vote(test_data:tuple, models:list):
        X_test, y_test = test_data
        pred_sum = np.array([0]*len(X_test))
        for model_str in models:
            # deserialize the model and predict
            model_bytes = model_str.encode("latin1")
            model = pickle.loads(model_bytes)
            pred_sum += model.predict(X_test)
        # prediction is set equal to the majority class
        n_models = len(models)
        threshold = np.ceil(n_models/2)
        preds = [int(s>=threshold) for s in pred_sum]
        print(f"Accuracy is: {accuracy_score(y_test,
            preds):.2f}")


    ### main flow ###
    # load and partition dataset
    partitioned_dataset = partition()
    train_data = partitioned_dataset["train"]
    test_data = partitioned_dataset["test"]
    # train in parallel 5 independent classifiers
    m1 = train(GaussianNB(), train_data)
    m2 = train(LogisticRegression(), train_data)
    m3 = train(DecisionTreeClassifier(), train_data)
    m4 = train(SVC(), train_data)
    m5 = train(KNeighborsClassifier(), train_data)
    # compute voting accuracy on test data
    vote(test_data, [m1, m2, m3, m4, m5])

# start DAG
ensemble_taskflow()
```

Listing 4.16: Ensemble learning in Airflow.

First, the DAG implementing the ensemble learning workflow has to be instantiated. For this purpose, the @dag decorator can be used to turn the ensemble_taskflow() function into a DAG generator. This annotation also allows users to specify several options, such as tags, and schedules for programmed executions. The workflow is composed of the following tasks, i.e., Python functions decorated via the @task annotation:

- *load*: It loads the *UCI breast cancer*[25] dataset using the *load_breast_cancer* function of *sklearn.datasets*. Data and targets are finally returned into a tuple, which will be JSON encoded and stored by XCom.
- *partition*: It receives the loaded dataset from the *load* task and partitions it into train and test sets. Afterward, it returns partitioned data as a dictionary. It is worth noting that, in this case, the multiple_outputs=True option is used in the task decorator to set the dictionary keys as XCom keys. It is useful to work with the XCom object given in output by the task as a dictionary with custom keys (more on this later).
- *train*: It receives training data and an instance of an sklearn estimator and performs the fitting operation. At this point, the fitted model should be returned, but sklearn estimators are not directly JSON serializable, which is required by XCom. To overcome this issue, the model is first serialized with the Pickle library into a sequence of bytes, which is then decoded into a JSON serializable string and returned as output.
- *vote*: This task receives test data and a list of fitted models. It first deserializes input models, using them for computing a list of predictions. Afterward, predictions are aggregated by voting to obtain the ensemble classification. In particular, as we dealt with a binary classification task using n base classifiers, with n being an odd number, final predictions are computed as follows. Let $S^t = \sum_{i=1}^{n} p_i^t$ be the sum of the predictions of all classifiers on the test sample t. So, the ensemble prediction for test sample t will be 1 if $S^t \geq \lceil n/2 \rceil$, or 0 otherwise.

After all tasks have been defined, the main flow is declared by composing tasks following the schema provided in Figure 4.13. In particular, we used five different classification models provided

---

[25]https://archive.ics.uci.edu/ml/datasets/Breast+Cancer+Wisconsin+ (Diagnostic).

by sklearn: a gaussian naive bayes, a logistic regression, a decision tree, a support vector machine, and a k-nearest neighbors classifier. A train task is created for each model, which runs in parallel. It is worth noting how, as explained in the definition of the partition task, the `multiple_outputs=True` option in the task decoration allows us to extract the train and test data from the XCom object given in output by the task, as the associated keys are set as XCom keys. Finally, the DAG is started by invoking the `ensemble_taskflow()` function.

4.4 Message Passing-Based Programming Tools

This section discusses the Message Passing Interface (MPI) as an example of a programming tool supporting the message-passing paradigm. A few concurrent programming languages, such as Smalltalk, Occam, Erlang, Scala, Rust, and Go, include message-passing mechanisms for process cooperation. However, MPI is the most used library for the implementation of message-based parallel and distributed applications on parallel and distributed computing systems. There are several open-source MPI implementations, such as MPICH, LAM/MPI, and Open MPI, which are largely used in languages such as Java, C, C++, Fortran, and Python.

4.4.1 *Message Passing Interface*

The MPI is a message-passing standard defined since 1992 by the MPI forum,[26] which is composed of many industrial and academic organizations. MPI is widely used by researchers and the industry for developing parallel and distributed applications in high-performance infrastructures. The MPI library includes point-to-point message passing, collective communications, group and communicator mechanisms, process topologies, environmental management, process creation and management, one-sided communications, extended collective operations, external interfaces, and I/O operations. Language bindings for C, Java, Fortran, Python, and R are defined. Although

---

[26] www.mpi-forum.org.

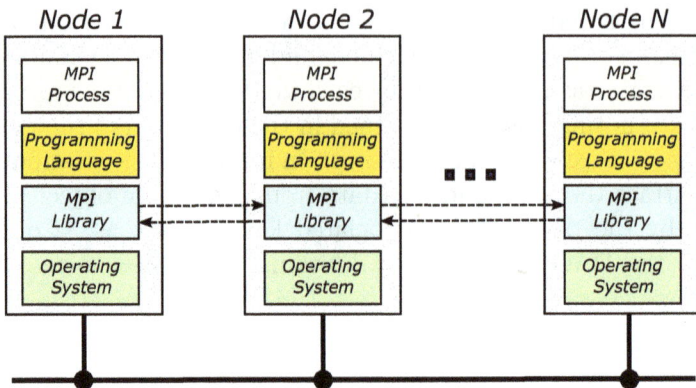

Fig. 4.14.   Levels of an MPI software architecture.

the MPI user community is medium in size, the project engages many contributors.

The first version MPI-1 released in May 1994 provided a rich collection of messaging primitives based on a set of eight basic functions that enable it to fully express parallel programs and other 129 advanced functions. An MPI-1 parallel program was composed of a set of similar processes running on different processors that use MPI functions for message passing (Figure 4.14). In the most recent versions, primitives for process initialization, creation, and management have been introduced. The latest version is MPI-4.1, which includes persistent collectives, partitioned communications, and new mechanisms for error handling.

4.4.1.1   *Basics*

According to the *Single Program Multiple Data* (SPMD) model, the MPI library is designed for exploiting *data parallelism* because all the MPI processes that compose a parallel program execute the same code on different data elements. Examples of MPI point-to-point communication primitives are:

- MPI_Send(buf, leng, type, rank, tag, comm);
- MPI_Recv(buf, leng, type, source, tag, comm, status).

MPI_Send and MPI_Recv calls operate as follows. Process P_1 packs up data to be sent into a buffer for process P_2. Then, the buffer is routed to the proper location. The location of the message is defined by the process rank. Process P_2 has to acknowledge that it wants to receive P_1's message by executing the receive function. Afterward, data are transmitted. Process P_1 is acknowledged that the data have been transferred and may execute the next operation.

The *leng* parameter contains the number of elements in the send buffer, *type* specifies the data type of each send buffer element, *rank* specifies the destination process, and *source* specifies the sender process. The *tag* is used to differentiate messages, in case P_1 sends multiple pieces of data to P_2. The *comm* parameter specifies the used communicator. The *status* argument for MPI_Recv provides information about the received message.

A communicator specifies the communication context for a message-passing operation. Each communication context provides a so-called "communication universe" composed of a set of processes that share this communication context. Messages are always received within the context they were sent, and messages sent in different contexts do not interfere. The MPI_COMM_WORLD default communicator is provided by the MPI. It allows communication with all processes that are accessible after MPI initialization and processes are identified by their ranks (numbers) in the group of MPI_COMM_WORLD.

A simple use of point-to-point communication is shown in the C programming example presented in Listing 4.17. In the example, the process with ID equal to 0 (myrank = 0) sends a data packet (a string) to the process with ID equal to 1 using the send operation MPI_Send. The location, length, and type of the send buffer are specified by the first three parameters of the MPI_Send operation. The message sent will contain 31 characters. In addition, the send operation associates an envelope with the packet. It specifies the packet destination and contains information that can be used by the receive operation to select a particular message. The last three parameters of the MPI_Send operation, along with the rank of the sender, specify the envelope for the packet sent. In particular, number 99 is chosen as tag for this message. Process 1 (myrank = 1) receives the text with the receive operation MPI_Recv. The packet to be received is selected according to the value of its envelope, and the text is stored in the receive buffer that will contain the string message in the memory of process 1. The first three parameters of the receive operation specify the location, length, and type of the receive buffer. The next three parameters, which include the tag, are used for selecting the incoming message. The last parameter is used to return information on the received message. Note that MPI_Comm_rank determines the rank (i.e., the ID) of the calling process in the communicator, and MPI_Init and MPI_Finalize are used to initialize and terminate the MPI execution environment, respectively.

```c
#include "mpi.h"
int main(int argc, char *argv[])
{
    char packet[50];
    int myrank;
    MPI_Status status;
    MPI_Init(&argc, &argv);
    MPI_Comm_rank(MPI_COMM_WORLD, &myrank);
    if (myrank == 0)    /* instructions of process with id 0 */
    {
      strcpy(packet,"Hello, this is my first message");
      MPI_Send(packet, strlen(packet)+1, MPI_CHAR, 1, 99,
          MPI_COMM_WORLD);
    }
    else
       if (myrank == 1) /* instructions of process with id 1 */
       {
         MPI_Recv(packet, 50, MPI_CHAR, 0, 99, MPI_COMM_WORLD, &
             status);
         printf("I received this packet:%s:\n", packet);
       }
    MPI.Finalize();
}
```

Listing 4.17: Simple message exchange through point-to-point communication in MPI.

MPI also provides a send–receive operation that combines in one operation the sending of a message to one destination and the receiving of another message from another process. The two processes (source and destination) are possibly the same.

```
MPI_SENDRECV(sendbuf, sendleng, sendtype, rank, sendtag,
recvbuf, recvleng, recvtype, source, recvtag, comm, status);
```

The send-receive operation is useful for implementing remote procedure calls (RPCs) and for executing a shift operation across a chain of processes.

Collective communication in MPI is implemented by the following primitives:

- `MPI_Bcast (inbuf, incnt, intype, root, comm);`
- `MPI_Scatter (inbuf, incnt, intype, inbuf, incnt, intype, root, comm);`
- `MPI_Gather (outbuf, outcnt, outype, inbuf, incnt, intype, root, comm);`
- `MPI_Reduce (inbuf, outbuf, count, type, op, root, comm).`

`MPI_Bcast` implements a broadcast communication that allows one process to send the same data to all processes in a communicator. While `MPI_Bcast` sends the same data to all processes, `MPI_Scatter` sends chunks of an array of data to different processes. `MPI_Gather` is the inverse of `MPI_Scatter`. Instead of spreading elements from one process to many processes, `MPI_Gather` takes elements from many processes and gathers them to one single process. Finally, similar to `MPI_Gather`, `MPI_Reduce` takes an array of input elements on each process and returns an array of output elements to the root process. The output elements contain the reduced result. MPI also includes a function dedicated to synchronizing processes: `MPI_Barrier(MPI_Comm comm)`. This function implements a barrier, and no processes in the communicator can pass the barrier until all of them call the function.

MPI provides a *low level of abstraction* for developing efficient and portable iterative parallel applications, even if performance may be limited by the communication latency between processors. MPI programmers cannot exploit any high-level construct and must manually cope with complex distributed programming issues, such as data exchange, the distribution of data across processors, synchronization, and deadlock. Those issues make it difficult to debug an application and to program tasks on end-user parallel applications, where higher-level languages are required to simplify the developer task. However, thanks to the efficiency arising from its low-level programming model, MPI is largely used and has been implemented on a very large set of parallel and sequential architectures, such as massively parallel processing systems, workstation networks, clusters, and grids. As mentioned before, MPI-1 did not make any provision for process creation, which was introduced later in the MPI-2 (Geist *et al.*, 1996) version through the *MPI_Comm_spawn* function that spawns a number of identical processes or the *MPI_Comm_spawn_multiple* function that spawns multiple processes, or the same process with multiple sets of arguments. The current version, MPI-4, provides extensions to better support hybrid programming models and fault tolerance.

4.4.1.2   *Programming example*

Here, we discuss an MPI application written in Java for counting characters in parallel through a set of processes reading from a text file. The example code is written in Java + Open MPI.[27] The syntax of MPI operations used here follows the Java syntax format. In particular, given an input file of M bytes and N processes, each process reads a chunk of $\frac{M}{N}$ bytes and counts each character in a private data structure. The master process, with rank equal to 0, receives the partial counts from all the processes within the group with the specified tag and aggregates them, as shown in Figure 4.15.

Listing 4.18 shows the application code. It uses the following primitives of MPI: (i) `Init` and `Finalize` to initialize and terminate the program, respectively; (ii) `bcast` to broadcast messages from the master to all the processes; and (iii) `send` and `recv` for point-to-point

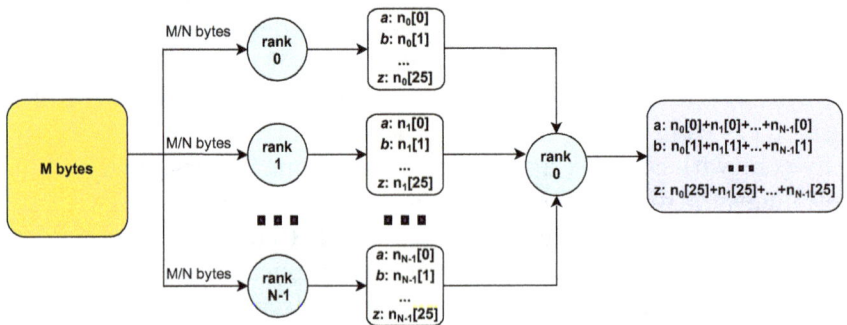

Fig. 4.15.   Architecture of the proposed MPI application.

---

[27]https://www.open-mpi.org/.

communication between master and the other processes. To run the application, the source code must be compiled with the mpijavac command and executed using the mpirun -N command, where N is the number of processes.

```java
static public void main(String[] args) throws MPIException,
    IOException {
    int tag = 42;
    // Initialize the message array for each process and
        result array for master
    int[] partial_counter = new int[26];
    int[] res = new int[26];
    int[] split_size = new int[1];
    String fileName = args[0];
    MPI.Init(args);
    Comm comm = MPI.COMM_WORLD;
    int rank = comm.getRank();
    int master_rank = 0;
    int ntasks = comm.getSize();
    if (rank == master_rank){ // The master computes the
        split size
        Path path = Paths.get(fileName);
        long bytes = Files.size(path);
        split_size[0] = (int) (bytes / ntasks);
    }
    // The split size is broadcast to all processes
    comm.bcast(split_size, split_size.length, MPI.INT, 0);
    int size = split_size[0];
    byte[] readBytes = new byte[size];
    try (InputStream inputStream = new FileInputStream(
        fileName)) {
        // Each process determines the chunk in which to
            work
        int start = rank * size;
        inputStream.skip(start);
        inputStream.read(readBytes, 0, size);
        // Count and store all the characters of the chunk
        for (byte b : readBytes) {
            char c = (char) b;
            if (Character.isLetter(c)){
                int index = (int) Character.toLowerCase(c) -
                    (int) 'a';
                partial_counter[index] += 1;
            }
        }
```

```
        // Each process communicates the partial counter to
           the master
        comm.send(partial_counter, partial_counter.length,
           MPI.INT, master_rank, tag);
    }
    if (rank == master_rank) {
        // The master aggregates the partial results
        for (int i = 0; i < ntasks; i++) {
            Status status = comm.recv(partial_counter,
                partial_counter.length, MPI.INT, MPI.
                ANY_SOURCE, tag);
            for (int j = 0; j < partial_counter.length; j++)
                res[j] += partial_counter[j];
        }
    }
    MPI.Finalize();
}
```

Listing 4.18: Parallel character count in MPI.

When the program starts, only the master process is executed. Following the MPI.Init method within the master process, $N-1$ additional processes are created to reach the number of parallel processes, N, indicated in the mpirun command. To identify a process, the MPI uses an integer ID, called *rank*, for each process, which is 0 for the master and is incremented each time a new process is created. In this way, the master can check the condition $rank == master\_rank$ to perform the following two operations: (i) establish the split size of a chunk for each process; and (ii) aggregate the partial character counts received by the processes. Communication is handled by the default communicator (i.e., *MPI.COMM_WORLD*), which groups all the processes to enable message exchange. Then, each process, including the master, continues to run distinct versions of the program. In particular, after receiving the split size broadcast by the master, the processes read the assigned data chunk, count the occurrences of each character, and store the result in a private structure (*partial_counter*). Finally, each process sends the partial counter results to the master in order to compute the final result.

4.5 BSP-Based Programming Tools

There are several frameworks and libraries that are based on the bulk synchronous parallel (BSP) model and are used for processing large-scale datasets, with a particular focus on efficient graph computation. Many of them are implementations of the Pregel (Malewicz *et al.*, 2010) model, proposed by Google in 2010, a BSP messaging abstraction aimed at expressing graph-parallel iterative algorithms. Among them, Apache Hama and Giraph are open-source BSP-based computing frameworks built on top of Apache Hadoop. They are designed to scale to large datasets in a distributed environment. Apache Flink offers built-in support for iterative BSP batch programming in the DataSet API, while also providing a vertex-centric graph API, namely Gelly, which exploits Flink's efficient iteration operators to support large-scale iterative graph processing. Apache Spark also provides its own optimized version of the Google Pregel graph processing system within the GraphX library. It relies on a graph-based high-level extension of Spark RDD APIs that allows programmers to exploit both primitives for graph computation and distributed data-parallel operations on Spark RDDs. In the following, this library is described in detail, showing how it can be leveraged for developing efficient graph-parallel applications.

4.5.1 *Spark GraphX*

A graph is a data structure composed of a set of vertices, also referred to as nodes, that are connected by edges. A graph can be (i) directed, if its edges have a defined orientation; (ii) cyclic, if there is at least one closed path; or (iii) weighted, if each edge is assigned a weight, which can represent an arbitrary property. Graphs are well suited to represent nonlinear relationships between objects, which has led to their application in several application domains, ranging from networking and social media analysis, to search engines and recommendation systems, employed by big companies, including Amazon, Netflix, and Google.

Although the high flexibility and expressiveness of graphs make them a powerful abstraction to enhance data analytics, data-parallel tools, and frameworks like Hadoop or Spark are not very well suited when dealing with graphs. Indeed, despite being designed for highly scalable and fault-tolerant data processing, they employ data abstractions that do not consider the graph structure, leading to excessive data movement and performance degradation. This entails the need for *ad hoc* solutions specifically designed for efficient graph-parallel computation. Among them, Spark *GraphX* provides a constantly growing set of graph operators, including property and structural ones, which can be used to perform map operations, create subgraphs, reverse the direction of the edges in a directed graph, or compute a masked version of an input graph. It also comes with some basic graph-based queries and algorithms for graph analytics, such as connected components identification, PageRank, and triangle counting. In addition, GraphX provides an optimized version of the Google Pregel graph processing system, proposed by Google in 2010 (Malewicz *et al.*, 2010), aimed at expressing graph-parallel iterative algorithms. The Pregel operator is a BSP messaging abstraction that performs a series of supersteps in which a vertex receives an aggregation of its inbound messages from the previous superstep, computes a new value for its property according to a task-specific logic, and finally sends messages to neighboring vertices in the next superstep.

4.5.1.1   *Graph abstraction*

At a high level, GraphX extends the Spark RDD by introducing a new *Graph* abstraction, a directed multigraph with properties attached to each vertex and edge. In particular, a graph contains two RDDs: one for edges and the other for vertices. This abstraction provides a unified interface to represent data, taking into account the underlying graph structure, while maintaining the efficiency of Spark RDDs. Indeed, it allows programmers to simultaneously exploit graph concepts, efficient primitives for graph computation, as well as distributed data-parallel operations typical of Spark.

GraphX also provides a further view of the underlying graph structure through the *EdgeTriplet* concept, which extends the information provided by the edge RDD by adding the properties of the source and destination vertices. In particular, an edge RDD only contains the property attached to each specific edge, along with the IDs of the source and destination vertices, i.e., a set of tuples (*src_id, dest_id, attr*). The properties of the source and destination vertices can be found in the RDD of vertices, which contains a tuple (*vertex_id, attr*) for each vertex of the graph. Therefore, the edge triplet, directly provides the complete information about two interconnected vertices, along with the attributes attached to them and to the edge that interconnects them, i.e., the tuple (*src_id, dest_id, src_attr, edge_attr, dst_attr*). This representation can be easily computed, starting with edge and vertex RDDs, as shown in Listing 4.19.

```
SELECT src.id, dst.id, src.attr, e.attr, dst.attr
FROM edges AS e LEFT JOIN vertices AS src, vertices AS dest
ON e.srcId = src.Id AND e.dstId = dst.Id
```

Listing 4.19: Derivation of the EdgeTriplet representation.

GraphX offers multiple options to create a graph from a set of vertices and edges in an RDD. In particular, a graph can be created from RDDs containing vertices and edges via the *Graph.apply* method or directly from an RDD of edges by using *Graph.fromEdges*. The *Graph.fromEdgeTuples* method enables the creation of graphs from edge tuples, i.e., an RDD containing pairs of vertices. GraphX also supports edge de-duplication and handles missing attributes in graph construction by means of default values.

In order to enable the development of scalable graph-parallel distributed applications, graphs generated in GraphX are partitioned. Particularly, a vertex-based approach for graph partitioning is exploited with the aim of lowering the communication and storage cost. The process consists of partitioning the graph along its vertices, which corresponds to assigning edges to machines and extending vertices along multiple machines.

The *Graph.partitionBy* operator allows users to set the specific edge assignment logic by selecting from a variety of partitioning algorithms, such as *2D partitioning* or other heuristics provided by GraphX.

4.5.1.2 *Pregel API*

In general terms, the characteristics of the vertices of a graph recursively depend on those of their neighbors, which makes graphs inherently recursive data structures. Due to this, many significant graph algorithms iteratively compute the characteristics of vertices until a fixed-point or convergence condition is met. These iterative algorithms have been expressed using a variety of graph-parallel abstractions. Among them, a variant of the *Pregel* API is exposed by GraphX.

The Pregel operator is a BSP message-passing abstraction in which the computation consists of a sequence of supersteps. During a Pregel superstep, the framework invokes a user-defined function (UDF) for each vertex, which specifies its behavior and runs in parallel. Each vertex can change its status and read the messages sent in the previous superstep or send new ones, which will be delivered in the next superstep. Communication usually takes place between a node and its neighborhood, although in theory, a message can be sent to any node with an identifier known to the sender. The topology of the graph serves as a constraint on computation in the Pregel API offered by GraphX and, as in the standard Pregel, the characteristics of vertices and edges are recalculated in an iterative manner until convergence is reached. Nonetheless, there are some differences that allow for substantially more efficient distributed graph-based computation. Specifically, in contrast to Pregel, GraphX performs message computation in parallel as a function of the edge triplet and has access to both the source and destination vertex properties. In addition, vertices can only send messages to their neighbors and those that do not get a message within a superstep are skipped. In the end, when there are no more messages to be processed, the Pregel operator stops iterating and returns the final state of the graph.

4.5.1.3   *Basics*

GraphX provides a rich set of functions and graph properties that can effectively support programmers in building efficient and concise graph-parallel applications. Given an input graph, useful information can be easily obtained, such as the number of edges (`val numEdges: Long`) and vertices (`val numVertices: Long`). The degree information can also be obtained as:

- `val inDegrees: VertexRDD[Int]`
- `val outDegrees: VertexRDD[Int]`
- `val Degrees: VertexRDD[Int]`

As stated in the previous section, graph abstraction provides a unified interface to deal with a graph, leveraging both the topological information and the efficiency of the RDD-based representation. For this reason, it can be represented as a collection of vertices, edges, and triplets:

- `val vertices: VertexRDD[VD]`
- `val edges: EdgeRDD[ED]`
- `val triplets: RDD[EdgeTriplet[VD, ED]`

GraphX also provides several mapping functions for transforming the attributes associated with vertices, edges, and triplets by specifying a user-defined map function:

- `def mapVertices[VD2](map: (VertexId, VD) => VD2): Graph[VD2, ED]`
- `def mapEdges[ED2](map: Edge[ED] => ED2): Graph[VD, ED2]`
- `def mapTriplets[ED2](map: EdgeTriplet[VD, ED] => ED2): Graph[VD, ED2]`

Among the many other useful functions provided by GraphX, the join operation is worth mentioning, which can be used to merge data from external RDDs with the graph. In particular, given an input RDD, the *joinVertices* operator joins the vertices of the graph that match with those contained in the input RDD. A new graph is then returned, in which the updated values of the joined vertices are computed by applying a user-defined map function, while non-matching vertices of the graph retain their original values.

```
def joinVertices[U](table: RDD[(VertexId, U)])(mapFunc: (VertexId,
                 VD, U) => VD): Graph[VD, ED]
```

A more general version of this function is *outerJoinVertices*, which applies the user-defined map function to all vertices of the graph. In this case, matching vertices are updated as usual, while the vertices that do not have a matching value in the input RDD are set to an *Option* type.

```
def outerJoinVertices[U, VD2](other: RDD[(VertexId, U)])(mapFunc: (
           VertexId, VD, Option[U]) => VD2): Graph[VD2, ED]
```

Optional values are useful to deal with situations in which a value can be either present or unset. In particular, in the Scala programming language, a *Option[T]* behaves like a container for zero or one element of type *T*, so it can be either *Some[T]* or a *None* object, which represents a missing value. When retrieving the value associated with an option type, a default value, to be used whether the Option is unset, can be specified using the `getOrElse(x)` method, where x is the default value.

In addition to the described functions, GraphX provides an implementation of the Pregel vertex-centric computation model, aimed at enabling the implementation of highly parallel large-scale graph applications. A programmer can use the Pregel-like BSP API through the following function.

```
def pregel [A]
      (initialMsg: A,
       maxIter: Int = Int.MaxValue,
       activeDir: EdgeDirection = EdgeDirection.Out)
      (vprog: (VertexId, VD, A) => VD,
       sendMsg: EdgeTriplet [VD, ED] => Iterator [(VertexId, A)],
       mergeMsg: (A, A) => A) : Graph [VD, ED]
```

This function takes two sets of input parameters: the first one specifies the input graph, the initial message (of a given type *A*), the number of iterations, and the edge direction; the second one expects three UDFs, as listed in the following:

- `vprog: (VertexId, VD, A) => VD`: It encodes the behavior of the vertex. This UDF is invoked on each vertex that receives a message and computes the updated vertex value.
- `sendMsg: EdgeTriplet [VD, ED] => Iterator [(VertexId, A)]`: This UDF is applied to the outgoing edges of vertices that received messages in the current iteration.
- `mergeMsg: (A, A) => A`: It specifies how two messages received by a vertex should be merged into a single message of the same type. This UDF must implement a commutative and associative function.

At the end of the computation, the resulting graph is returned in the output.

4.5.1.4 *Programming example*

In this section, we give an example of the Pregel operator in action by showing how it can be leveraged to implement the famous *PageRank* algorithm, which is used by Google Search to rank website pages. This algorithm was first introduced in 1998 by Lawrence Page and Sergey Brin in their paper about the initial prototype of the Google search engine (Brin and Page, 1998). The basic idea behind PageRank is that more important websites are likely to receive more links from other websites. Thus, it is aimed at estimating how important a website is by determining the quality of links pointing to that website. More specifically, PageRank models the process that leads a user to a given page. The user generally lands on that page through a sequence of random links, i.e., by following a path connecting multiple pages. However, the user can eventually stop clicking on outgoing links and make a random hop by searching for a different URL that is not directly reachable from the current page. The probability that a user will continue clicking on outgoing links is the *damping factor*, generally set to $d = 0.85$, while the random hop probability is $1 - d = 0.15$. Formally, this process is expressed by the following equation:

$$PR(p_i) = \frac{1-d}{N} + d \left(\sum_{p_j \in M(p_i)} \frac{PR(p_j)}{L(p_j)} \right)$$

The PageRank of page p_i represents the likelihood of a user, located on a page p_j, to land on p_i. It is given by the sum of two probabilities:

- Given that the probability of the user to stop clicking on links present in p_j and make a random hop is $1 - d$, and assuming that the probability of landing on each of the N available pages is equally distributed, this term expresses the probability to reach p_i through a random hop.
- Given d the probability of the user to follow an outgoing link, $M(p_i)$ the set of pages linking to p_i, and $L(p_j)$ the number links in each page $p_j \in M(p_i)$, this terms gives the probability of the user to land on page p_i by following an existing link in p_j, assuming that the probability of the user to follow each link in p_j is equally distributed.

Listing 4.20 shows the implementation of the PageRank algorithm using the Pregel API provided by GraphX.

```scala
object Pagerank {

  //Create the Spark session
  val spark = SparkSession
    .builder
    .master("local")
    .appName("Spark-GraphX-PageRank")
    .getOrCreate()
  val sc: SparkContext = spark.sparkContext

  // Build an example graph
  val vertices: RDD[(VertexId, Double)] = sc.parallelize(Seq(1, 11)
    .map(x => (x.asInstanceOf[Long], x.asInstanceOf[Double])))
  val edges: RDD[Edge[PartitionID]] = sc.parallelize(Seq(Edge(2L,
    3L, 1), Edge(3L, 2L, 1), Edge(4L, 2L, 1), Edge(4L, 1L, 1),
    Edge(5L, 4L, 1), Edge(5L, 6L, 1), Edge(5L, 2L, 1), Edge(6L, 5L,
      1), Edge(6L, 2L, 1), Edge(7L, 2L, 1), Edge(7L, 5L, 1),
    Edge(8L, 2L, 1), Edge(8L, 5L, 1), Edge(9L, 2L, 1), Edge(9L, 5L,
      1), Edge(10L, 5L, 1), Edge(11L, 5L, 1)))
  val graph: Graph[Double, PartitionID] = Graph(vertices, edges)
  val numVertices = graph.numVertices
  val startGraph: Graph[Double, Double] = graph
    // Associate the degree with each vertex
    .outerJoinVertices(graph.outDegrees) {
      (vid, vdata, deg) => deg.getOrElse(0)
    }
    // Set the weight on the edges based on the out-degree
    .mapTriplets(e => 1.0 / e.srcAttr)
    // Set the vertex attributes to an initial guess
    .mapVertices((id, attr) => 1)

  // Set the damping factor
  val d = 0.85

  // Define Pregel's UDFs for PageRank
  def vertexProgram(id: VertexId, attr: Double, msgSum: Double):
      Double = (1-d)/numVertices + d * msgSum

  def sendMessage(edge: EdgeTriplet[Double, Double]): Iterator[(
      VertexId, Double)] = Iterator((edge.dstId, edge.srcAttr *
      edge.attr))

  def messageCombiner(a: Double, b: Double): Double = a + b

  def main(args: Array[String]) = {
    // Execute Pregel for a fixed number of iterations.
```

```
val finalGraph = Pregel(graph = startGraph, initialMsg = 0.0,
    maxIterations = 50)(vprog = vertexProgram, sendMsg =
    sendMessage, mergeMsg = messageCombiner)
// Normalize to sum 1 (it is required to deal with sink nodes)
val rankSum = finalGraph.vertices.values.sum()
val rankGraph = finalGraph.mapVertices((id, rank) => rank /
    rankSum)
// Show top 5 nodes ordered by decreasing PageRank
rankGraph.vertices.top(5)(Ordering.by(_._2)).foreach(println)
  }
}
```

Listing 4.20: Pagerank implementation in GraphX using the Pregel API.

The code starts by creating the *SparkSession*, passing typical parameters such as the master or the application name. Then, the *SparkContext* is derived from the session. Subsequently, an example graph is created, which is composed of two RDDs containing edges and vertices. This graph is then prepared for the PageRank computation by setting the weight of each edge equal to the out-degree of the source vertex and by setting the initial PageRank value of each vertex to an initial guess equal to 1.0. Afterward, the damping factor is set to $d = 0.85$, and the three UDFs are defined. In particular: (i) the *vertexProgram* UDF implements the aforementioned PageRank formula by taking as input the sum of normalized PageRank contributions of the in-neighboring pages; (ii) the *sendMessage* UDF encodes how a page shares its PageRank equally among its out-neighbors; (iii) the *messageCombiner* UDF implements the sum of the contributions that will be used in the *vertexProgram* to update the rank of the page receiving those contributions. Finally, a main function is shown in which the Pregel operator is executed for 50 iterations, and after the BSP computation, the top five nodes by PageRank are printed. Note that, in order to deal with the presence of sink nodes, the output of the Pregel operator is further renormalized to ensure a probability distribution among all the available pages.

The output of the proposed example is summarized in Figure 4.16, where each circle represents a page annotated with the computed PageRank. Vertices are shown in different gradations of green according to the final computed PageRank.

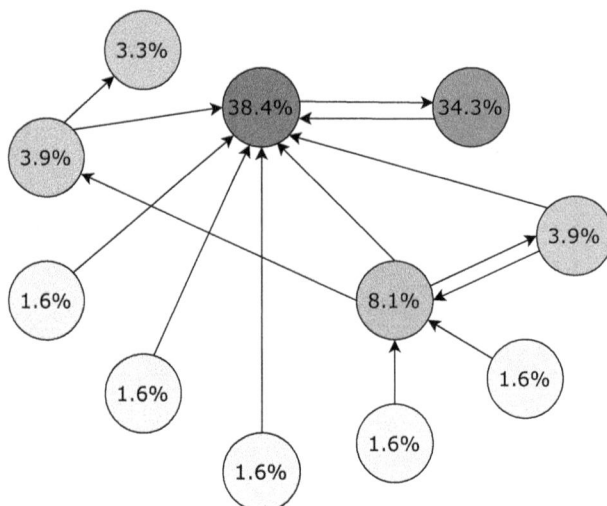

Fig. 4.16.   The output of PageRank in the proposed example.

4.6   SQL-Like Programming Tools

In this section, we discuss some frameworks that support the SQL-like programming model. These systems attempt to combine Hadoop's efficacy and querying capabilities with the SQL-like language's ease of use in order to enable the development of simple and efficient data analysis applications.

Apache Hive[28], a data warehouse software built on top of Hadoop for reading, writing, and managing data in large-scale infrastructures, is one of the most commonly used systems in this context. *Apache Pig*[29] is another Hadoop-based framework that exploits a SQL-like language for executing data flow applications in large-scale infrastructures.

Slightly different is *Apache Impala* (Kornacker *et al.*, 2015), a massively parallel query engine that runs on Hadoop data processing environments, providing low latency and high concurrency for analytic queries on Hadoop while offering an RDBMS-like experience.

In the following, we discuss in detail the basics of Apache Hive and Apache Pig, as representatives of SQL-like programming tools.

---

[28]https://hive.apache.org/.
[29]https://pig.apache.org/.

4.6.1 *Apache Hive*

Hive is a Hadoop-based data warehouse system, which enables users to write queries using a SQL-like declarative language, called HiveQL, which are then compiled into MapReduce jobs and run on Hadoop. It can be considered as a SQL engine capable of automatically compiling a SQL-like query into a set of MapReduce jobs that are run on a Hadoop cluster, with additional features for data and metadata management. The reasons behind the development of such a system are based on the fact that, although MapReduce is a very flexible programming paradigm, it is too low level for routine data analysis tasks. The decision to use SQL as the higher-level language for Hive was driven by the aim of facilitating the transition for analysts and data scientists from traditional relational databases to Hadoop's distributed processing, leveraging the familiarity and widespread adoption of SQL. In fact, Hive supports a large portion of the SQL standard, as well as several extensions designed to make interactions with the underlying Hadoop platform easier. For example, SQL databases organize data into tables, which are then divided into typed columns. Hive follows a similar approach, but unlike traditional databases, it does not require defining the table structure before importing data. Instead, it allows projecting tabular structure into the underlying data during query execution. This feature is known as schema-on-read, which means that data are checked against the schema when a query is performed on it. Hive also allows the execution of the same query on different portions of data, supporting *data parallelism*. Moreover, it provides a *high level of abstraction* because a programmer can develop a data processing application by using HiveQL, which relies on traditional concepts of relational databases.

Today, Apache Hive is commonly used by data analysts for *data querying* and *reporting* on large datasets. It is supported by a large user community and is adopted by several big companies, such as Facebook, Netflix, Yahoo!, and Airbnb. For example, Netflix uses Hive for *ad hoc* queries and analytics.

4.6.1.1 *Main concepts*

Hive offers a comprehensive range of Data Definition Language (DDL) and Data Manipulation Language (DML) operations. It supports essential functionalities such as table creation, alteration, browsing, and deletion. Furthermore, it also facilitates data loading, insertion, update, deletion, and merging on the file system. These capabilities make Hive a versatile option that can replace existing data warehouse infrastructures, acting as an adapter between the Hadoop platform and the large ecosystem of data analysis tools built on top of relational databases (e.g., ETL and business intelligence applications). However, it is important to note that Hive is specifically designed for online analytical processing (OLAP) rather than online transaction processing (OLTP). Additionally, unlike relational databases such as SQL Server, Hive does not provide real-time access to data.

The abstractions used by the HiveQL language rely on traditional concepts of relational databases (e.g., table, row, column). In addition, Hive provides three different types of functions for data manipulation, which are user-defined functions (UDFs), user-defined aggregate functions (UDAFs), and user-defined table-generating functions (UDTFs). UDFs operate on a single row and produce a single row as output, such as *length*, *round*, and *factorial*. UDAFs operate on multiple rows and produce a single row as output, such as *sum*, *average*, and *count*. Finally, UDTFs operate on single rows and produce multiple rows as output, such as *explode*, *posexplode*, and *inline*. Such functions make it really easy to write custom ones in different languages, such as Java or Python.

4.6.1.2 *Architecture*

The architecture of Hive comprises the following components, also shown in Figure 4.17 (Huai *et al.*, 2014):

- *User Interface* (UI): It serves as the entry point for users to submit queries and perform other operations within the system via a web UI or a command line interface (CLI).
- *Driver*: This component receives queries and implements session handles, providing execute and fetch APIs modeled on JDBC/ODBC interfaces.
- *Compiler*: It is responsible for parsing queries, performing semantic analysis on query expressions, and generating an execution plan. In particular, it converts the query to an abstract syntax tree (AST), and then, after checking for compilation errors, the

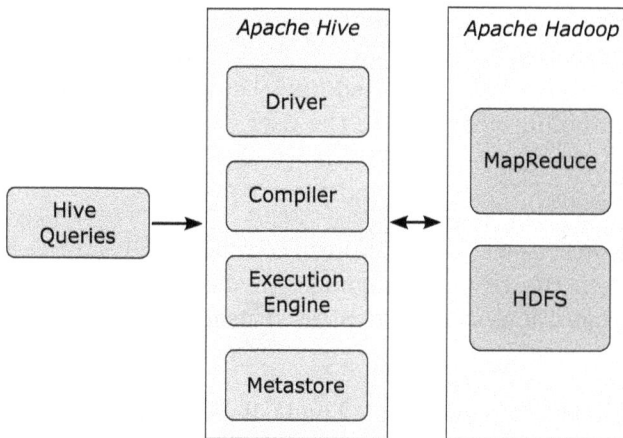

Fig. 4.17. The architecture of Apache Hive.

AST is converted into a DAG. The compiler utilizes table and partition metadata obtained from the Metastore.

- *Metastore*: This component uses an RDBMS to store structural information about various tables and partitions in the warehouse, such as metadata of persistent relational entities and how they are mapped onto HDFS. This information includes also column details, column types, serializers, deserializers for data reading and writing, and the corresponding HDFS file locations.

- *Execution engine*: It runs the execution plan generated by the compiler in the form of a DAG. The execution engine manages dependencies between stages of the DAG and executes them on the appropriate system components.

- *HDFS*: It is the underlying distributed file system used for data storage in Hadoop.

The typical flow of a HiveQL query through a Hive architecture is as follows. The UI initiates the execution interface to the Driver, which creates a session handle for the query and sends it to the Compiler for the generation of an execution plan. The Compiler obtains necessary metadata from the Metastore, using it for type-checking expressions and applying a set of optimizations to the AST (e.g., pruning partitions based on query predicates). The operator tree is then converted into a DAG of multiple MapReduce jobs, which are submitted to the underlying MapReduce engine for evaluation.

A serialization–deserialization library, named *SerDe*, is used to serialize and deserialize data for a specific file format when users write files to HDFS. After all the MapReduce jobs are completed, the Driver will return the query results to the user.

4.6.1.3 *Basics*

As already discussed, Hive supports common DDL and DML operations through the HiveQL language. In this section, we discuss how the main operations can be expressed using the language proposed by Hive.

HiveQL DDL statements: The DDL is a syntax for creating and altering database objects, such as tables and indices. As an example, to create a table through the HiveQL language, programmers can use the following statement:

```
CREATE [REMOTE] (SCHEMA|DATABASE) [IF NOT EXISTS] database_name
    [LOCATION hdfs_path] [ROW FORMAT row_format] [FIELDS
                    TERMINATED BY char];
```

As discussed in Section 4.6.1.2 on architectural aspects, Hive keeps the schema of the tables in the Metastore component, which is used to store all the information about tables and partitions. The default Metastore is the Apache Derby Database,[30] which can be explicitly expressed using the SCHEMA (or the equivalent DATABASE) option.

With a similar syntax, Hive allows the user to modify and remove rows or delete an entire table. Other common DDL statements are detailed in Table 4.3.

HiveQL DML statements: The DML is a syntax used for inserting, deleting, and updating data in a database. As an example, for loading data into a table through the HiveQL language, programmers can use the following statement:

```
LOAD DATA [LOCAL] INPATH 'filepath' [OVERWRITE] INTO TABLE
    tablename [PARTITION (partcol1=val1, partcol2=val2,"...");]
```

---

[30]https://db.apache.org/derby/.

Table 4.3. Common DDL statements in HiveQL.

DDL statement	Meaning
SHOW	Show databases, tables, and properties.
ALTER	Modify entries in an existing table.
DESCRIBE	Describe table columns.
TRUNCATE	Delete table rows.
DELETE	Delete table data.

When importing data into tables, Hive does not perform any transformations, whereas load operations are just copy/move operations that move data into Hive table locations. It is also worth noting that the results of a query can be inserted into tables using a syntax similar to the previous one. The OVERWRITE option will overwrite any existing data in the table or partition.

```
INSERT OVERWRITE TABLE tablename [IF NOT EXISTS]
    select_statement FROM from_statement;
```

Other ways to modify data in Hive are UPDATE and DELETE, which can be used through the following statements:

```
UPDATE tablename SET column = value [, column = value "..."]
    [WHERE expression];
DELETE FROM tablename [WHERE expression];
```

4.6.1.4 *Programming example*

The proposed application shows how Hive can be employed to store data about the ratings of users for a set of movies and to perform some queries on these data using HiveQL.

As a first step, we need to create a table in which data will be stored. This is shown in Listing 4.21, where a table is created by specifying its schema, i.e., four columns named *userid*, *movieid*, *rating*, and *timestamp*.

```
CREATE TABLE data (
  userid INT,
  movieid INT,
  rating INT,
  timestamp DATE)
```

```
ROW FORMAT DELIMITED
FIELDS TERMINATED BY '\t';
```

Listing 4.21: Creating a Hive table.

Once the table is created, data are retrieved from a text file stored in the local file system (see Listing 4.22), overwriting any contents of the table. It is worth noting that by removing the LOCAL keyword, Hive will look for the file in HDFS.

```
LOAD DATA LOCAL INPATH '<path>/data'
OVERWRITE INTO TABLE data;
```

Listing 4.22: Loading data into a Hive table.

Afterward, as shown in Listing 4.23, we can perform some statistical analyses, such as counting the number of rows (i.e., the number of user ratings) in the table by using the built-in COUNT function.

```
SELECT COUNT(*)
FROM data;
```

Listing 4.23: Counting rows.

By using traditional SQL clauses, such as *SELECT, FROM, GROUP BY*, and *ORDER BY*, it is possible to perform some simple queries, such as finding the most highly rated movies (see Listing 4.24).

```
SELECT movieid, COUNT(rating) AS num_ratings
FROM data
GROUP BY movieid
ORDER BY num_ratings DESC
```

Listing 4.24: Most highly rated movies.

Moreover, Hive also allows users to perform more complex data analyses using other programming languages. For example, we can define a Python script, named *week_mapper.py*, which performs some processing steps on the rows, as shown in Listing 4.25. Specifically, this script maps the timestamp of each movie review to the week of the year it was produced and aggregates the results. The *isocalendar()* function returns a triple with (year, week, day); therefore, *isocalendar()*[1] returns the week number.

```
import sys
import datetime

for line in sys.stdin:
  line = line.strip()
  userid, movieid, rating, timestamp = line.split('\t')
  week = datetime.datetime.fromtimestamp(float(timestamp)).
    isocalendar()[1]
  print '\t'.join([userid, movieid, rating, str(week)])
```

Listing 4.25: Python script.

The Python script discussed in Listing 4.25 can easily be incorporated into HiveQL statements and used as a function, as shown in Listing 4.26. In particular, through the *TRANSFORM* clause, users can insert their own custom functions into the data stream to execute a custom script. By default, columns will be transformed into strings and delimited by tabs before being sent to the user script. This is why in Listing 4.25, we split a row by tab and then recompose a new row again using the tab. Finally, a COUNT query is performed by grouping the new data by week of the year.

```
add FILE week_mapper.py;

INSERT OVERWRITE TABLE data_new
SELECT TRANSFORM
      (userid, movieid, rating, timestamp) USING
      'python week_mapper.py'
  AS (userid, movieid, rating, week)
FROM data;

SELECT week, COUNT(*)
FROM data_new
GROUP BY week;
```

Listing 4.26: Hive query.

4.6.2 *Apache Pig*

Apache Pig is a high-level dataflow framework for executing MapReduce programs on Hadoop by using a language called Pig

Latin. In 2008, when the first version of Pig was released, the success of internet companies became heavily reliant on their capacity to analyze vast amounts of data gathered on a daily basis. As a result, *ad hoc* analysis gained increasing importance. Although parallel database systems were available, they were excessively expensive, leading to the wide diffusion of procedural programming models and systems for data analysis, such as MapReduce with Hadoop. Despite its widespread use, the MapReduce programming model was characterized by a low level of abstraction, demanding highly specialized programmers and making maintenance and reuse difficult. To cope with these issues, Pig was proposed to bridge the gap between the high-level declarative querying of SQL and the low-level procedural style of the MapReduce programming model (Gates *et al.*, 2009). Similarly to Hive (see Section 4.6.1), queries in Pig are written using a custom language, named Pig Latin, and are then converted into execution plans that are performed as MapReduce jobs on Hadoop.

The Pig programming system allows for composing high-level data manipulation operations using a SQL-like style (e.g., parallel programming operations, such as FOREACH, FLATTEN, and COGROUP) while retaining the main characteristics, data types, and workloads of MapReduce. In addition, Pig exploits a multi-query execution system to process an entire script or a batch of statements at once. Thus, it supports both *data parallelism*, which is exploited by partitioning data into chunks and processing them in parallel, and *task parallelism*, when multiple queries run in parallel on the same data.

For these reasons, Pig is commonly used for developing *data querying*, simple *data analysis* and extract, transform, load (ETL) applications, gathering data from several sources, such as streams, HDFS, or files. The companies and organizations that use Pig in production include LinkedIn, PayPal, and Mendeley. Thanks to the Pig Latin scripting language, Pig provides a *medium level of abstraction*, which means that, compared to other systems such as Hadoop, Pig developers are not required to write complex and lengthy codes.

4.6.2.1 *Main concepts*

Data model: Pig provides a nested data model, which allows for handling complex and non-normalized data (Olston *et al.*, 2008b). It supports scalar types, such as int, long, double, chararray (i.e., string), and bytearray types. Moreover, it provides three complex data models, namely map, tuple, and bag:

- A *map* is an associative array, where the key is a string and the value can be any type.
- A *tuple* is an ordered list of data elements, also called *fields*, where each field is a piece of data. The elements of a tuple can be of any type, thus allowing nested complex types.
- A *bag* is a collection of tuples, similar to a relational database. Thus, tuples in a bag correspond to the rows in a table, although, unlike a relational table, Pig bags do not require that each tuple contain the same number of fields. A bag is also identified as a relation.

A Pig script can use a large set of operators to ease the development of common tasks on data, such as LOAD, FILTER, JOIN, and SORT. Pig scripts can be invoked by applications written in many other programming languages besides Pig Latin (e.g., Java, Python, and JavaScript), and it can exploit custom UDFs for advanced analytics.

Query optimization: Each Pig script is translated into a set of MapReduce jobs that are automatically optimized by the Pig engine using several built-in optimization rules, such as reducing unused statements or applying filters during data loading. The optimization of the dataflow can be *logical* or *physical* (Olston *et al.*, 2008a).

Logical optimizations rearrange the logical dataflow graph submitted by the user, generating a new graph that is semantically equivalent to the original one but can be evaluated in a more efficient way. Conversely, physical optimizations concern how the logical dataflow graph is translated into a physical execution plan, such as a series of MapReduce jobs. Pig creates a logical plan for each bag defined by the user. When the logical plan is built, no processing occurs, but it starts only when the user invokes a STORE command on a bag. At that point, the logical plan for that bag is transformed into a physical plan, which is then performed. This lazy execution is advantageous since it allows for in-memory pipelining and other optimizations, such as filter reordering over many Pig Latin commands.

4.6.2.2   *Architecture*

From an architectural perspective, Pig consists of four major components, as also shown in Figure 4.18:

- A *parser*, which handles all Pig Latin statements, checking for syntax and data types errors. It produces a DAG as output, which represents the logical operators of the scripts as nodes and data flow as edges.
- An *optimizer*, which applies optimization operations on the DAG generated by the parser for improving query speed, such as split, merge, projection, pushdown, transform, and reorder. For example, pushdown and projection omit unnecessary data or columns, reducing the amount of data to be processed.
- A *compiler*, which generates a sequence of MapReduce jobs, starting from the output of the optimizer. This process includes other optimizations, such as rearranging the execution order.
- An *execution engine*, which executes MapReduce jobs generated by the compiler on the Hadoop runtime. The output can be shown on screen using the DUMP command or saved in HDFS using the STORE function.

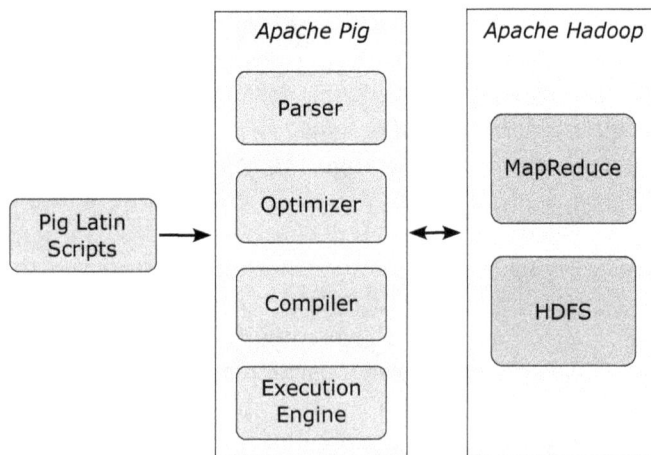

Fig. 4.18.   The architecture of Apache Pig.

4.6.2.3 *Basics*

As already discussed in Section 4.6.2.1, Pig Latin statements are expressed using bags. A bag is a collection of tuples (i.e., ordered sets of fields), which can be created by using native data types supported by Pig, both simple types (e.g., int, long, float, double, chararray, and boolean) or complex ones (i.e., tuples, maps, or nested bags), or by loading data from the file system. It is important to note that relations are referred to by a name (or alias), which is assigned as part of the Pig Latin statement. An example of a Pig Latin statement that loads data related to students (i.e., names and ages) from a file is shown as follows:

```
A = LOAD 'file' USING PigStorage() AS (name:chararray, age:int);
```

Other common Pig Latin statements that express relational operators are listed in the following (note that a tuple is enclosed in parentheses ( ), a bag is enclosed in curly brackets { }, and a map is enclosed in square brackets [ ]):

- FILTER, which selects tuples from a relation based on some condition. The syntax is:

```
alias = FILTER alias  BY expression;
```

- JOIN (inner or outer), which performs an inner/outer join of two or more relations based on common field values. The syntax is:

```
alias = JOIN alias | left-alias BY { expression };
```

- FOREACH, which generates data transformations based on columns of data. The syntax is:

```
alias  = FOREACH { block | nested_block };
```

Usually, the use of the FOREACH operation is coupled with the GENERATE operation, which enables working with columns of data (instead of the FILTER operation that works with rows). The syntax is:

```
X = FOREACH A GENERATE f1;
```

- STORE, which stores or saves results to the file system. The syntax is:

```
STORE alias INTO 'directory' [USING function];
```

Moreover, since Pig supports the definition of UDFs by the programmer, it allows for registering a JAR file to be used in the script and for assigning aliases to UDFs.

```
REGISTER path;
DEFINE alias {function | ['command' [input] [output] [stderr]]};
```

4.6.2.4 *Programming example*

The application discussed here implements a dictionary-based sentiment analyzer using Pig. Given a dictionary of words associated with positive or negative sentiment, the sentiment of a text (e.g., a sentence, a review, a tweet, or a comment) is calculated by summing up the scores of positive and negative words in the text and by calculating the average rating, as shown in Figure 4.19.

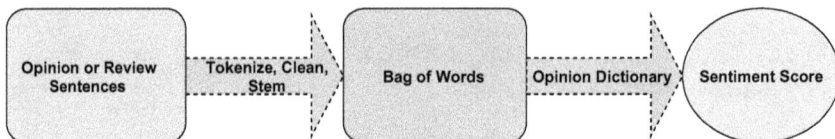

Fig. 4.19.   Structure of the sentiment analyzer application.

Since Pig does not provide a built-in library for sentiment analysis, the system exploits external dictionaries to associate words with their sentiments and determine the semantic orientation of the opinion words (Kumar and Sebastian, 2012; Belcastro *et al.*, 2020).

As already discussed, developers can include advanced analytics in a script by defining UDFs. For example, the *PROCESS* UDF in Listing 4.27 is aimed at processing a tuple by removing punctuation as a preprocessing step. Other functionalities, if required, can be added to the *exec* method, which is implemented in Java.

```
public class Processing extends EvalFunc<String> {
    @Override
    public String exec(Tuple tuple) throws IOException {
        if (tuple == null || tuple.size() == 0 ||
            tuple.get(0) == null)
            return null;
```

```
String str = (String) tuple.get(0);
// Remove punctuation and, if required, apply
    lemmatization, stemming and other
String clean = str.toLowerCase().replaceAll
    ("\\p{Punct}", "");
...
return clean;
    }
}
```

Listing 4.27: Java UDF processing.

The code in Pig Latin for the sentiment analyzer is described in Listing 4.28. First, the UDF defined in Java is registered, and an alias, named *PROCESS*, is defined to use it as a built-in function (see Listing 4.28).

```
REGISTER PigUDF.jar;
DEFINE PROCESS main.Processing;
```

Listing 4.28: Registering a Java UDF.

Afterward, data related to text reviews are loaded from HDFS as a *CSV* file delimited by tab, and similarly, the dictionary of word sentiments is loaded. When loading data from a file, the user can specify its schema through named columns, as shown in Listing 4.29.

```
-- Load data from HDFS
reviews = LOAD 'hdfs://hostname:port/pigdata/reviews.csv' USING
    PigStorage ('\t') AS (id:int, text:chararray);
-- Load the dictionary of word sentiment
dictionary = LOAD 'hdfs://hostname:port/pigdata/dictionary.txt'
    USING PigStorage('\t') AS (word:chararray,rating:int);
```

Listing 4.29: Loading review data and word sentiment dictionary.

Once data are loaded, each row is tokenized and processed according to the function defined in Listing 4.27. In particular, as shown in Listing 4.30, through the FOREACH operator, each row of the input data is first processed using the previously registered UDF and then tokenized, producing an array of tokens as output. This array is subsequently flattened through the built-in FLATTEN operator. The GENERATE operator is used in cooperation with the FOREACH to produce as output triples in the form of ⟨review_id, text, word⟩, which are stored in a bag named *words*.

```
-- Tokenize and process the text of each review
words = FOREACH reviews GENERATE id,text, FLATTEN(TOKENIZE(
    PROCESS(text))) AS word;
```

Listing 4.30: Tokenization and preprocessing.

Specifically, the code in Listing 4.31 first identifies all matches among the words of a review and the words of the dictionary by joining the intermediate bag created above and the words in the dictionary. The outcomes are stored in a bag named *matches*, which is then iterated to assign the score to each word. The result of this operation, stored in a bag named *matches_rating*, is a triple in the form ⟨review_id, text, rating⟩.

```
-- Join each word with the dictionary and assign a score
    sentiment
matches = JOIN words BY word, dictionary BY word;
matches_rating = FOREACH matches GENERATE words::id AS id,
    words::text AS text, dictionary::rating AS rate;
```

Listing 4.31: Joining words with the dictionary and assigning a score sentiment.

The pair ⟨review_id, text⟩ is used to perform a group by operation and collect all ratings found in the dictionary (i.e, the bag named *group_rating*), as shown in Listing 4.32. After the grouping, for each review, the built-in AVG operator is used to aggregate all word ratings, and the final rating of a review is calculated as the average of the scores of its tokens. Finally, the output bag *avg_ratings* is stored in a file on the underlying HDFS file system.

```
-- Group and compute the average rating for a review
group_rating = GROUP matches_rating BY(id,text);
avg_ratings = FOREACH group_rating GENERATE group,AVG($1.$2) AS
    rate;
-- Store the results
STORE avg_ratings INTO 'ratings' USING PigStorage(',','-schema');
```

Listing 4.32: Calculating average rating and storing results to HDFS.

4.7 PGAS-Based Programming Tools

In this section, we discuss some tools that support the partitioned global address space (PGAS) programming model, which represents a trade-off between distributed- and shared-memory programming models. Specifically, PGAS has been designed for implementing a global memory address space that is logically partitioned and portions of it are local to single processes. The main goal of the PGAS model is to limit data exchange and isolate failures in very large-scale systems.

The asynchronous PGAS (APGAS) model is a variant of PGAS that supports both local and remote asynchronous task creation. Unlike PGAS, APGAS does not require that all processes run on similar hardware and supports the dynamic spawning of multiple tasks. In fact, multiple threads can be active simultaneously in a place, using either local or remote data. In addition, it does not require that all the places in a computation must be homogeneous.

In recent years, several PGAS-based languages have been proposed for supporting efficient computing on large-scale distributed systems, such as DASH (Fuerlinger *et al.*, 2016), X10 (Charles *et al.*, 2005), Chapel (Deitz *et al.*, 2006), pPython (Byun *et al.*, 2022), and UPC (Consortium UPC, 2005).

In the following, we discuss in detail the UPC++ library, an implementation of UPC developed at Berkeley Lab. It is a well-maintained and complete solution for developing PGAS-based applications. Moreover, UPC++ developers distribute ready-to-use Docker containers and detailed manuals that make it easy to learn and use.

4.7.1 *UPC++*

UPC++ (Zheng *et al.*, 2014) is a library based on C++ that provides classes and functions aimed at supporting APGAS programming. Besides accessing the local memory, as in standard C++, in the APGAS model, each thread, referred to as *rank*, has access to a global address space allocated in shared segments that are distributed over the ranks. This memory model, as depicted in Figure 4.20, makes UPC++ suitable for writing parallel programs that run efficiently and scale well on distributed-memory parallel computers made up of hundreds of thousands of cores.

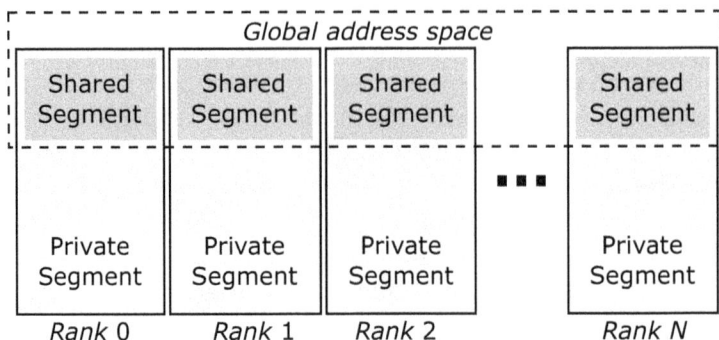

Fig. 4.20.   APGAS memory model.

UPC++ uses the *global-address space networking* (GASNet) middleware, a language-independent layer that provides network-independent communication primitives, including remote memory access (RMA) and active messages (AM). Due to its high performance and portability, GASNet has been used to implement parallel global address space languages and libraries, such as the aforementioned UPC++, Legion, and Chapel. In UPC++, all remote memory access operations are by default asynchronous, and interfaces are designed to be composable and similar to those used in conventional C++. Its main programming abstractions, discussed later in this section, include global pointers, RPCs, futures, and shared objects.

UPC++ provides a low level of abstraction for developing large-scale iterative parallel applications, facilitating fine-grained control over parallelism and efficient utilization of computing resources. However, performance may be hindered by the extensive use of communication, and application debug can be challenging. It presents a high level of *verbosity* due to the absence of high-level constructs, which forces programmers to manually handle some challenges in distributed programming, such as data exchange and synchronization. UPC++ exploits *data parallelism* during execution since input data are distributed across multiple ranks and processed in parallel.

4.7.1.1   *Basics*

All UPC++ programs include two fundamental operations:

- `upcxx::init()`, which initializes the UPC++ runtime and must be called before any UPC++ features are used.

- `upcxx::finalize()`, which closes down the UPC++ runtime and prevents any UPC++ features to be used after it.

A UPC++ program runs with a fixed number of threads, referred to as *ranks*, each executing one copy of the program. Each rank has an associated identifier, i.e., an integer number ranging from 0 to $N - 1$, with N denoting the number of ranks, which can be accessed via the `upcxx::rank_me()` operation.

Asynchronous computation: Computation in a UPC++ program can be divided across different ranks, allow them to perform their operations in parallel. The final result is then collected from one of the ranks, which means that there is a sync point where the result of each rank is awaited. In order to improve the level of parallelism, asynchronous computation is leveraged in UPC++, which allows for overlapping communication and computation, actively exploiting waiting time. This is achieved through the use of *future* objects, which are UPC++ constructs characterized by a value and a state, indicating whether the value is available (*ready*) or not. As an example, let us consider the `upcxx::allreduce` operation, defined as follows:

```
template<typename T, typename BinaryOp>
upcxx::future<T> upcxx::allreduce(T &&value, BinaryOp &&op,
                   upcxx::team &team = upcxx::world());
```

This operation performs a global reduction of a value of type T across all ranks by applying a binary function (e.g., a sum). In addition, the use of *teams* is supported. Teams are ordered sets of ranks to which collective operations can be applied. Currently, in UPC++, the only supported team is `upxx::world()`, which comprises all the ranks. The return type of the reduce operation shown above is *future⟨T⟩*, which allows for asynchronous computation. In particular, a rank waiting for a result does not block until that value is available but can perform other unrelated operations in the meantime. This logic is implemented by the method `wait()` of `upcxx::future`, which internally checks for the state of the future object, looping until it completes and becomes *ready*.

Shared objects: According to the APGAS model, UPC++ allows programmers to work with objects shared across different ranks. In particular, a *shared object* is allocated within a shared memory segment and is accessible through a global pointer, which is defined as upcxx::global_ptr<T> gptr.

As depicted in Figure 4.21, UPC++ defines two different memory areas in the global address space, in which objects can be allocated through one of the following approaches:

- Using upcxx::new_<T>, a new object of type T is allocated in the shared segment of the current rank. Each rank can reference this object through a private global pointer to its local shared segment.
- Using the standard C++ new keyword, a typical dynamic allocation is performed in the rank's private local memory.

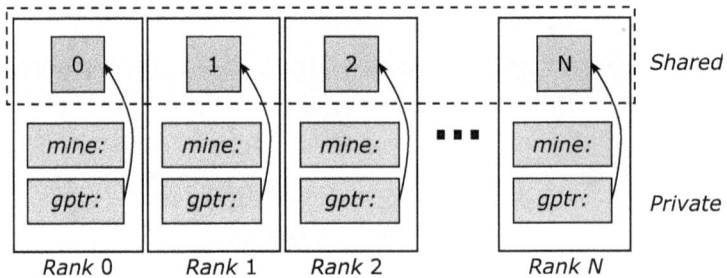

Fig. 4.21. UPC++ global pointers.

4.7.1.2 *Programming example*

This section presents an example, included in the UPC++ library, which shows how to compute a Monte Carlo estimate of π by leveraging APGAS programming. The Monte Carlo method for estimating π works by generating a large number of points in a 2D plane, whose coordinates (x, y) are uniformly distributed random variables in the interval $[0, 1]$. Among all the randomly generated points, the method counts how many of them fall within the sector of the circumference with unit radius centered at $(0, 0)$, i.e., all the points which satisfy the following equation: $x^2 + y^2 \leq 1$. This point generation process is depicted in Figure 4.22.

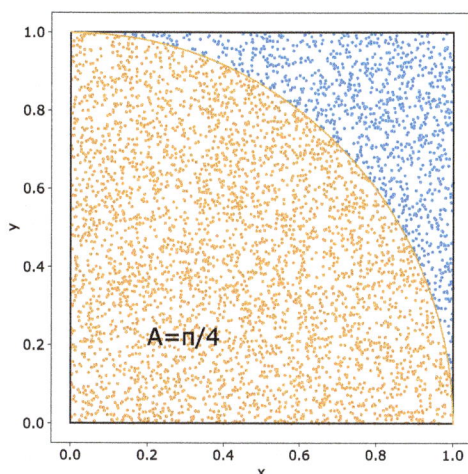

Fig. 4.22. Monte Carlo estimation of π.

Let N be the total number of points generated and M be the number of points that fall within the circular sector. Then, given that the areas of the unit square and the section are, respectively, $A_q = 1$ and $A_s = \frac{\pi}{4}$, the value of π can be estimated as follows:

$$\frac{M}{N} = \frac{A_s}{A_q} = \frac{\pi}{4} \longrightarrow \pi = 4 \cdot \frac{M}{N}$$

Listing 4.33 shows an implementation in UPC++ of the estimation process described above. As described in the previous section, the UPC++ program starts by invoking the upcxx::init() method, which sets up the UPC++ runtime. Then, each rank, whose identifier is given by the upcxx::rank_me() method, initializes a local random generator, which is used to compute the coordinates of the different trials. Each rank executes in parallel and performs 100,000 trails by generating a random point in the unit square, with coordinates (x, y), $x \sim U[0, 1]$, $y \sim U[0, 1]$, by counting how many of them fall within the circular sector of unit radius. This value, i.e., the number of hits, is stored by each rank in its local variable *my_hits*. Once each rank has generated its own trials, all results are collected by rank 0 (more on this later), which determines the total number of hits through the *reduce_to_rank0* method and stores them in the *hits* variable. Finally, given the total number of ranks, achieved via upcxx::rank_n(), the overall number of generated points (*trials*) can be computed, and the estimate of π can be derived as 4 * *hits* / *trials*.

```cpp
#include <iostream>
#include <cstdlib>
#include <random>
#include <upcxx/upcxx.hpp>

using namespace std;

int64_t hit()
{
    // generate random point coordinates
    double x = static_cast<double>(rand()) / RAND_MAX;
    double y = static_cast<double>(rand()) / RAND_MAX;
    // check whether the point falls within the circular
        sector
    if (x*x + y*y <= 1.0) return 1;
    else return 0;
}

int main(int argc, char **argv)
{
    upcxx::init();
    // each rank gets its own copy of local variables
    int64_t my_hits = 0;
    // the number of trials to run on each rank
    int my_trials = 100000;
    // initialize a random number generator for each rank
    srand(upcxx::rank_me());
    // local computation
    for (int i = 0; i < my_trials; i++) {
        my_hits += hit();
    }
    // rank 0 collects and sum the hits of all ranks
    int64_t hits = reduce_to_rank0(my_hits);
    // rank 0 prints the final estimate
    if (upcxx::rank_me() == 0) {
        int64_t trials = upcxx::rank_n() * my_trials;
        cout << "pi estimate: " << 4.0 * hits / trials <<
            endl;
    }
    upcxx::finalize();
    return 0;
}
```

Listing 4.33: Monte Carlo estimation of π in UPC++.

What remains to be clarified is how the local results of each rank are used by rank 0 to determine the overall number of hits, from which the final estimate of π is derived. UPC++ provides different ways to perform this operation, among which the most straightforward is to use the *allreduce* method, as shown in the following:

```
upcxx::allreduce(my_hits, plus<int>()).wait()
```

In this case, the collective reduction is computed by applying the *plus* function (i.e., a sum) over all local values of *my_hits*. This operation is asynchronous and returns a future of type int, so the final result has to be awaited through the *wait* method.

A different way to determine the overall number of hits is shown in Listing 4.34, which uses UPC++ global pointers based on the APGAS shared-memory model.

```
int64_t reduce_to_rank0(int64_t my_hits)
{
    // Rank 0 creates an array to store all of the incoming
        values
    upcxx::global_ptr<int64_t> all_hits_ptr = nullptr;
    if (upcxx::rank_me() == 0) {
        all_hits_ptr = upcxx::new_array<int64_t>(upcxx::
            rank_n());
    }
    // Rank 0 broadcasts the array global pointer to all
        ranks
    all_hits_ptr = upcxx::broadcast(all_hits_ptr, 0).wait();
    // All ranks offset the pointer of the array by their
        rank id
    upcxx::global_ptr<int64_t> my_hits_ptr = all_hits_ptr +
        upcxx::rank_me();
    // All rank put their local hits value into the shared
        array
    upcxx::rput(my_hits, my_hits_ptr).wait();
    // wait for all insertions to be completed
    upcxx::barrier();
    // rank 0 performs the reduce-by-sum operation
    int64_t hits = 0;
    if (upcxx::rank_me() == 0) {
        // get a local pointer to the shared array
        int64_t *local_hits_ptrs = all_hits_ptr.local();
        // sum all the values stored in the array
```

```
        for (int i = 0; i < upcxx::rank_n(); i++) {
            hits += local_hits_ptrs[i];
        }
        upcxx::delete_array(all_hits_ptr);
    }
    return hits;
}
```

Listing 4.34: Determining the total number of hits by using global pointers.

As a first step, rank 0 allocates a global pointer *all_hits_ptr* to an array of integer numbers via the upcxx::new_array function. The array size is set equal to the number of ranks, obtained as upcxx::rank_n(), since it will be used to store all hits values from remote ranks. The global pointer is then broadcast to all ranks, which will update a specific position of the array, based on their rank id obtained as upcxx::rank_me(). Particularly, the rank id is used to offset the global pointer so that the ith rank will insert the value of its own local *my_hits* variable into the ith position of the shared array. The insertion of the local hits value into the global shared array is performed by each rank using the upcxx::rput (*remote put*) function, which initiates a one-way communication to transfer to the remote process in an asynchronous fashion, so that no coordination is needed with the destination process. After passing upcxx::barrier, which acts as a sync point for ensuring that all remote transfers have been completed, rank 0 uses the upcxx::global_ptr<T>::local function to get a local version of the global pointer and sum up all the values inserted by remote ranks. Finally, it deallocates the shared array using the upcxx::delete_array function and returns the reduced result stored in the *hits* variable.

Chapter 5

Comparing Programming Tools

This chapter discusses and compares the programming tools presented in the previous chapter. A set of functional and non-functional properties is used to compare the frameworks. This comparison will be used to explain how each tool may support developers in programming big data applications.

5.1 Toward a Tool Analysis

Organizations and researchers in the area of big data are looking for ever more efficient tools to process and derive insights from large amounts of data. To meet different needs, a wide variety of programming tools designed for big data analytics are available today. This chapter aims at presenting a comprehensive comparison of the different programming tools introduced in the previous chapter. To narrow it down and make the comparison more meaningful, we also carry out a comprehensive analysis through application examples, focusing on four major classes of applications: batch, streaming, data querying, and graph-based. The goal of this example-driven comparison is to help the reader gain insight into the relative strengths and weaknesses of the major tools for specific use cases.

The chapter is organized as follows. Section 5.2 summarizes and classifies the main features of the discussed programming tools, including their diffusion, community support, advantages, and disadvantages. It aims to help developers choose between such tools based on several other factors, such as budget, the type of parallelism, the level of abstraction, verbosity in writing code, and the main classes of applications. Section 5.3 compares the various big data tools by evaluating their strengths and limitations through application examples. The comparison also takes into account additional factors such as performance, scalability, usability, and suitability.

5.2   Comparative Analysis of the System Features

This section summarizes and classifies the main features of all the discussed programming tools, their diffusion, advantages, and disadvantages. Some of these systems share features, making it difficult for programmers to choose between them. The choice can depend on several factors, such as budget (e.g., often high-level services are easy to use but more expensive than low-level solutions), the type of parallelism, data format, data source, the amount of data, and performance. Indeed, given a specific big data analysis task, it can be implemented using different programming models and systems.

The comparative analysis carried out in this section may help scholars and developers choose the best system based on their programming skills, parallel model, budget, application domain, and support provided by the community of both users and developers.

5.2.1   *System features*

Table 5.1 summarizes the features of the systems according to their programming model, type of parallelism, level of abstraction, verbosity in writing code, and main classes of applications.

The *level of abstraction* refers to the capability of a system to hide low-level programming details, thus allowing developers to focus on the problem logic. A high level of abstraction makes it easy to build applications but hard to compile them to efficient code. On the other hand, a low level of abstraction makes it hard to build applications but easy to implement them efficiently (Skillicorn and Talia, 1998). For comparison purposes, we distinguish three levels of abstraction:

- *Low*: Frameworks in this category provide powerful APIs and primitives that require distributed programming skills, resulting in a high development effort. They also require a low-level understanding of the system, including working with files in distributed

Table 5.1. Features of the systems.

System	Programming model	Type of parallelism	Level of abstraction	Verbosity	Main class of applications
Hadoop	MapReduce	Data	Low	High	Batch processing
Spark	Workflow BSP (Pregel API) SQL-like (Spark SQL)	Data/Task	Medium	Low	Batch and stream processing, graph analysis, data querying, iterative parallel applications
Storm	Workflow	Data/Task	Medium	Medium	Stream processing
Airflow	Workflow	Data/Task	High	Low	Pipeline-based applications
MPI	Message passing	Data	Low	High	Iterative parallel applications
Hive	SQL-like	Data	High	Low	Data querying and reporting
Pig	SQL-like	Data/Task	Medium	Low	Data querying and analysis
UPC++	PGAS-based	Data	Low	High	Iterative parallel applications

environments (Wadkar *et al.*, 2014). However, the code efficiency is high because it can be fully tuned. Examples of such frameworks include Hadoop, Message Passing Interface (MPI), and UPC++.

- *Medium*: Such systems allow developers to implement parallel and distributed applications using a limited number of constructs. They require some programming skills, but the development effort is lower than that of frameworks with a low level of abstraction. Examples include Spark, Storm, and Pig.

- *High*: This class includes systems requiring limited programming skills, enabling developers to rapidly build data analytics applications through simple visual interfaces or high-level scripts. At this level, though program development effort is low, the code efficiency at run time is also low because executable generation is more complex and code mapping is not direct. Hive and Airflow fall into this category.

The *type of parallelism* describes the way in which a programming model or system expresses parallel operations and how its runtime supports the execution of concurrent operations on multiple nodes or processors. For comparison purposes, we distinguish two types of parallelism:

- *Data parallelism*: It is achieved when the same code is executed in parallel on different data elements. Data parallelism is also known as single instruction multiple data (SIMD), which is a class in Flynn's taxonomy for classifying parallel computation (Flynn, 1972). This category includes Hadoop, Spark, Storm, MPI, UPC++, Hive, and Pig. Such systems are designed to automatically manage large input data, split them into chunks, and process them in parallel on different computing nodes.

- *Task parallelism*: It is achieved when different tasks that compose applications run in parallel. The presence of data dependencies can limit the benefits of this kind of parallelism. Such parallelism can be defined in the following two ways: *explicitly*, where a programmer defines dependencies among tasks through explicit instructions, and *implicitly*, where the system analyzes the input/output of tasks to understand dependencies among them. This form of parallelism is exploited by Spark, Storm, Pig, and

Airflow. Such systems enable the parallel execution of independent tasks without any data dependency.

As regards *verbosity*, systems can be classified as follows in view of the discussed programming examples:

- *High*: Systems in this category require a large number of lines of code and use many instructions/calls to build even a simple application. As a result, writing applications using these systems can be complex and time-consuming (Verma *et al.*, 2016). As an example, Hadoop is included in this category since a MapReduce application in Hadoop requires the definitions of mapper, reducer, and job.
- *Medium*: It includes systems that require implementing specific interfaces and methods to codify an application or to use specific constructs. For example, Storm requires implementing the interfaces for spouts and bolts and overriding methods such as *nextTuple* and *declareOutputFields*.
- *Low*: This category includes frameworks such as Spark, Airflow, Hive, and Pig. Writing code in these systems is usually compact because programmers are not forced to use specific constructs, and when needed, only a few lines of code are required. They usually provide an easy-to-use style of programming (e.g., HiveQL and Pig Latin) or may allow building workflows directly from a set of scripts (e.g., Airflow).

Finally, systems can support different *classes of applications*, as follows:

- *Batch*: Applications designed to process and analyze large volumes of data in a non-interactive mode.
- *Stream*: Applications designed to process and analyze data collected in real time.
- *Graph*: Applications designed to process and analyze data that are interconnected in complex networks or graph structures.
- *Data querying*: Applications designed to provide fast and efficient access to large volumes of data using query languages and search tools.

Furthermore, sometimes the following additional application classes are considered:

- *Iterative parallel*: Applications composed of repetitive tasks running in parallel.
- *Pipeline-based*: Applications composed of a sequence of stages where the output of one stage is fed as input to the next stage.

Programmers can decide to exploit some *general-purpose systems*, such as Hadoop, Spark, MPI, and UPC++, or systems designed for specific application domains. For example, the GraphX library available in Spark provides a version of the Google Pregel graph processing system, which can be effectively used to develop *graph processing* applications. Besides this, Hive and Pig can be leveraged for *data querying*, and Storm for *real-time stream processing*.

5.2.2   *System diffusion*

Table 5.2 summarizes the diffusion and popularity of each system from the user and developer perspectives.

As for *diffusion*, we classified the systems by considering the following parameters:

- *User community*, which refers to the diffusion of a system in terms of the number of people who actively use it. To compare the systems, we examined the number of relevant questions asked on Stack Overflow. In particular, the total number of questions was used as the main indicator of user interest, while the average number of questions per week was used to better capture the latest trends in user adoption. Based on this, we found that Spark had the largest user community, followed by Hadoop and Hive, and then MPI, Airflow, Pig, and Storm.
- *GitHub stars*, as an indicator of the *popularity* and reusability of the systems. In fact, there exists a strong relationship between system popularity and the user-perceived quality of that system and, therefore, its reuse. In particular, Papamichail *et al.* (2016) demonstrated a strong positive correlation between the number of stars and the number of forks by analyzing the 100 most popular Java repositories on GitHub. The most popular and reused system is Spark, followed by Airflow, Hadoop, Storm, Hive, MPI, and Pig.

Table 5.2. Diffusion and popularity of the systems.

System	User community size (weekly)	GitHub stars	API support	GitHub commits	Adopters
Hadoop	Large: 44.2k (16)	13.3k	Java, C, C++, Ruby, Groovy, Perl, Python	26.5k	Yahoo!, IBM, Amazon
Spark	Very large: 79.6k (139)	35.2k	Scala, Python, Java, R	36k	eBay, Amazon, Alibaba, CERN
Storm	Small: 2.5k (—)	6.4k	Clojure, Java, Python, Ruby	10.5k	Twitter, Groupon, Spotify
Airflow	Medium: 9.5k (50)	29.4k	Python	19k	Adobe, Onefootball
MPI	Medium: 6.8k (11)	1.7k	Java, Fortran, C, C++, Perl, Python	33.6k	Amazon WS, AMD, Cisco, Facebook
Hive	Large: 21.7k (27)	4.7k	HiveQL	16.6k	Facebook, Netflix, Yahoo!, AirBnB
Pig	Small: 5.2k (—)	656	PigLatin	3.7k	LinkedIn, PayPal, Mendeley
UPC++	Very small: 23 (—)	—	C++	—	NERSC, LANL

Source: The data were accessed from Stack Overflow and GitHub in March 2023.

- *API support*, which indicates the set of programming languages available to develop applications with that system. Systems such as Hadoop, Spark, Storm, Airflow, and MPI can accommodate developers with different programming skills and backgrounds thanks to the provision of APIs in popular languages (specifically, Python and Java, according to the PYPL index,[1] accessed March 2023). Conversely, frameworks such as Pig or Hive may require learning new languages, such as PigLatin or HiveQL.
- *GitHub commits*, as an indicator of the size of the developer community that contributes to the development and maintenance of code. We referred to the number of commits on official GitHub repositories to grasp the interest of developers in fixing errors/bugs

---

[1]https://pypl.github.io/PYPL.html.

and introducing new features. As seen in the diffusion among users, Spark showcased the most involvement from contributors, followed by MPI, which attracted the interest of programmers working on a wide range of applications, and Hadoop. Airflow, Hive, Storm, and Pig follow in order.

- *Main adopters*, which refer to the major companies and research institutions that actively engage with a particular system, incorporating it into their operations and research activities. The adoption serves as a testament to the system's reliability, functionality, and potential for widespread use. Big IT companies, such as IBM, Amazon, Twitter, Facebook, Netflix, and PayPal, widely employ some of the discussed frameworks, such as Hadoop, Spark, and MPI, affirming their relevance and impact in the industry. The adoption by these companies not only highlights the capabilities of the frameworks but also their importance as tools for achieving organizational goals. Conversely, other frameworks, such as UPC++, are mainly used by the research community.

5.2.3   *Advantages and disadvantages*

Table 5.3 summarizes the main advantages and disadvantages of the described systems when used for programming big data analysis applications. The pros and cons are outlined starting with features that emerged during the description of each system and discussed in the context of applications and code snippets, presented later.

The advantages of each system are related to the specific features it offers compared to related systems in terms of functionality, support for different libraries, and integration with other frameworks. For example, Spark is the only system that leverages an in-memory computing model, enabling the design of efficient data-intensive applications. Furthermore, as will become apparent in the following sections from the comparison with special-purpose frameworks, such as Storm and Pig, Spark is highly flexible and can be leveraged for a wide range of application domains, as it provides libraries for stream and graph processing, machine learning, and structured data analysis. Among the other workflow-based frameworks, Airflow is the most user-friendly, allowing anyone with sufficient Python knowledge to deploy, monitor, schedule, and manage a workflow through a web application.

Table 5.3.   Advantages and disadvantages of the systems.

System	Advantages	Disadvantages
Hadoop	Fault tolerance, low cost, very large open-source community	Verbosity, limited to batch processing, small file handling issues inefficient for iterative applications
Spark	In-memory computing, ease-of-use and flexibility, libraries for graph analytics, scalable machine learning support	No automatic optimization process, small file handling issues high memory consumption
Storm	Multi-language support, low-latency response time	Message ordering not guaranteed
Airflow	Support for Python script orchestration, workflow flexibility and control, extensibility and integrability	Workflow versioning is absent, lack of extensive documentation
MPI	Efficiency, portability, shared or distributed memory	Hard to debug, bottleneck in network communication
Hive	Large distributed datasets querying, SQL-like language, UDFs for advanced data analysis	OLAP-only support, real-time data access not supported, writing complex data analytics functions can be challenging
Pig	High-level procedural language, UDFs for advanced data analysis, easy learning and development	Small community, hard to tune performance
UPC++	High efficiency and scalability, support for APGAS programming	Hard to debug, extensive use of communication

On the other hand, system disadvantages are mainly related to deficiencies, weaknesses, costs, and limitations in the use of a given system. For example, Hadoop is not well suited for iterative applications but excels in batch processing. The main disadvantage of using Storm for real-time data stream computations is the lack of message ordering guarantees. MPI and UPC++ are generally efficient but can be hard to debug due to their low-level programming models. Hive is well suited for large distributed data querying; however, it does not support OLTP operations. Finally, Pig offers an easy-to-use programming interface for data analysis applications but debugging is complex.

5.3   Comparative Analysis through Application Examples

This section presents a comparative analysis of the different systems for big data, examining their strengths and limitations also through the lens of application examples. We aim to provide insights into the performance, scalability, usability, and suitability of different frameworks developed for handling big data challenges.

5.3.1   *Batch application: Apache Spark vs. Apache Hadoop*

In this section, we present an application for automatically discovering user mobility patterns from geotagged Flickr posts generated in the city of Rome. More in detail, the application aims at discovering the most frequent user trajectories across some specific locations or areas that are of interest for our analysis, commonly referred to as points of interest (PoIs). Specifically, a PoI is a location that is considered useful or interesting, such as a tourist attraction or business location. Since information on a PoI is generally limited to an address or GPS coordinates, it is hard to match trajectories with PoIs. For this reason, it is often useful to define the so-called regions of interest (RoIs) that represent the boundaries of the PoIs' areas (Belcastro *et al.*, 2018). In such a way, a trajectory can be defined as a sequence of RoIs, representing a movement pattern over time. A frequent trajectory is a sequence of RoIs that are frequently visited by users.

As illustrated in Figure 5.1, the workflow of the proposed application is composed of different steps. In particular, after collecting a set of geotagged posts from Flickr, we apply some preprocessing to filter out the geotagged posts and map each geotagged post to an RoI.

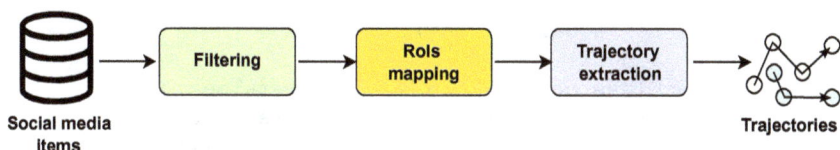

Fig. 5.1. Workflow of the trajectory mining application.

Finally, we apply trajectory mining to extract frequent mobility patterns in user trajectories across RoIs, aiming to better understand how people move in the city of Rome.

GPS sensors and devices are commonly used to collect data for trajectory analysis, as they provide regular updates on the device's location. However, social media data can also be used for this purpose (Cesario *et al.*, 2017). In fact, social media posts are often geotagged, which means they contain information about the post's location or other metadata that can be used to infer the user's position when the post was created. Specifically, a social media item may include the following fields (Belcastro *et al.*, 2021a):

- a textual description,
- a set of keywords associated with the post,
- a pair of latitude and longitude that represents the coordinates where the post was created,
- an ID that identifies the user who created the post,
- a timestamp that indicates the date of creation of the post.

5.3.1.1 *Spark implementation*

We first discuss the implementation of the application in Apache Spark using the Scala language. Listing 5.1 defines two classes, namely *SingleTrajectory* and *UserTrajectory*, to represent a trajectory (i.e., a pair $\langle PoI, timestamp \rangle$) and a user (i.e., a pair $\langle user\_id, timestamp \rangle$), respectively. These classes define the *equals* and *hashCode* methods that will be used by the subsequent procedures when comparing trajectories and users.

```
class SingleTrajectory(var poi: String = "", var daytime:
    String = "") {
  override def hashCode(): Int =
    31 * poi.## + daytime.##
```

```scala
  override def equals(o: Any): Boolean = {
    if (o == null) return false
    if (getClass ne o.getClass) return false
    val other = o.asInstanceOf[SingleTrajectory]
    if(poi.equals(other.poi) && daytime.equals(other.
        daytime))
      return true
    false
  }
}

class UserTrajectory (var username: String = "",
  var daytime: String = ""){

  override def hashCode(): Int =
    31 * username.## + daytime.##

  override def equals(o: Any): Boolean = {
    if (o == null) return false
    if (getClass ne o.getClass) return false
    val other = o.asInstanceOf[UserTrajectory]
    if(username.equals(other.username) && daytime.equals
        (other.daytime))
      return true
    false
  }
}
```

Listing 5.1: SingleTrajectory and UserTrajectory classes for representing a trajectory and a user, respectively.

The *.##* notation is equivalent to the *.hashCode* function, except for numeric types and null. For numeric values, the hash value returned by *.##* is consistent with value equality, meaning that if two instances compare as true, they will produce the same hash value with *.##*. However, when a null value is encountered, the *.##* operator returns the hash code, while *null.hashCode* throws a *NullPointerException*.

Then, in Listing 5.2, we define the *main* method that comprises the steps described in Figure 5.1.

```scala
def main(args: Array[String]): Unit = {
  val dataset = "FlickrRomeSample.json"
  val kmlPath = "rome.kml"
  val spark = SparkSession
```

```
    .builder
    .appName ("TrajectoryMining")
    .master ("local[*]")
    .getOrCreate()
  val stringShapeMap = KMLUtils.lookupFromKml (kmlPath).
    asScala
  var df = spark.read.json (dataset)
  df = filterFlickrDataframe (df)
  df = df.filter (r => filterIsGPSValid (r) &&
    filterIsInRome (r))
  val trajectories = computeTrajectoryUsingFPGrowth (df,
    stringShapeMap)
  trajectories._1.foreach (println)
  trajectories._2.foreach (println)
  spark.close()
}
```

Listing 5.2: Main method.

The main method reads the JSON input dataset into a Spark DataFrame and applies a series of functions to filter out data points that are not relevant to the analysis, such as those with invalid GPS coordinates or those not located in Rome. Next, the code uses the *KMLUtils* class to read a Keyhole Markup Language (KML) file containing the coordinates of some PoIs in Rome and convert it into a Scala map where the keys are the names of the PoIs (e.g., *colosseum*, *vaticanmuseums*, and *mausoleumofhadrian*) and the associated values are the coordinates that define the boundaries of that area as a polygon. The program then computes user trajectories using an FPGrowth algorithm and returns the frequent itemsets and the association rules that represent the trajectories mined. The *KMLUtils* class defines a utility method, namely *lookupFromKml*, that converts the *KML* file of Rome's PoIs to the previously described map using the Java API for KML to enable the convenient and easy use of KML in Java environments.

The code for the filtering step is described in Listing 5.3. In particular, we define three filters:

- a filter for selecting only the interesting columns from the Flickr dataset for the analysis,
- a filter that maintains only the data with valid GPS coordinates (i.e., whose longitudes and latitudes are correctly defined),
- a filter for maintaining only posts that were published in Rome.

```
def filterFlickrDataframe(dataframe: DataFrame): DataFrame
    = {
    dataframe.select("geoData.latitude", "geoData.
        longitude", "geoData.accuracy", "owner.id",
        "dateTaken")
}

def filterIsGPSValid(r: Row): Boolean = {
    r.getAs[Double]("longitude") > 0 && r.getAs[Double]
        ("latitude") > 0
}

def filterIsInRome(r: Row): Boolean = {
    val p = GeoUtils.getPoint(r.getAs[Double]("longitude"),
        r.getAs[Double]("latitude"))
    GeoUtils.isContained(p, romeShape)
}
```

Listing 5.3: Filtering methods.

The *filterIsInRome* function exploits a utility class, namely *GeoUtils*, that provides a set of utility methods for interacting with geospatial data, such as converting a pair of longitude and latitude data to a point or checking if a PoI is contained in another PoI. The class is based on the Spatial4j library, which is a general-purpose geospatial Java library that provides common shapes implementable in Euclidean and geodesic (surface of a sphere) world models, enables distance calculations, and allows for reading and writing shapes in formats such as WKT and GeoJSON.

After having filtered the input dataset, we can extract association rules using a frequent pattern mining algorithm. In particular, we used the FPGrowth algorithm, which is a popular algorithm for mining frequent itemsets in transactional databases proposed by Han *et al.* (2000). The FPGrowth algorithm uses a compressed data structure called the frequent pattern tree (FP-tree) to represent the transactional database, which allows for the efficient mining of frequent itemsets by scanning the tree and using a recursive process to construct conditional patterns. The algorithm consists of two phases: building the FP-tree and mining frequent itemsets. The first phase involves scanning the transactional database once and constructing the FP-tree by identifying the frequent items in the transactions and organizing them into a tree structure based on their frequency. The second phase involves recursively mining the frequent itemsets from the tree by identifying the frequent itemsets at each level in a bottom-up direction and building a conditional pattern base for each frequent itemset until all frequent itemsets have been mined.

To adapt the algorithm for trajectory mining we define a *mapCreateTrajectory* method that, given a row of the input dataset and the set of PoIs extracted from the KML file, returns a tuple consisting of the user ID and the PoI he/she visited in the form of the previously defined ⟨*UserTrajectory, SingleTrajectory*⟩. The code is shown in Listing 5.4.

```scala
def mapCreateTrajectory(row: Row, shapeMap: mutable.Map
    [String, String]): (UserTrajectory,
    Set[SingleTrajectory]) = {
  val lat = row.getDouble(0)
  val lon = row.getDouble(1)
  val username = row.getString(3)
  val d = row.getAs[String](4).substring(0, 12)
  val point = GeoUtils.getPoint(lon, lat)
  var arr: Array[SingleTrajectory]
      = Array[SingleTrajectory]()
  for ((place, polygon) <- shapeMap) {
    val pol = GeoUtils.getPolygonFromString(polygon)
    if (GeoUtils.isContained(point, pol)) {
      val trajectory = new SingleTrajectory(place, d)
      arr = arr :+ trajectory
    }
  }
  val ut = new UserTrajectory(username, d)
  (ut, arr.toSet)
}
```

Listing 5.4: Method for extracting a user and the PoI visited by them.

The first step of the *computeTrajectoryUsingFPGrowth* method, whose implementation is shown in Listing 5.5, is to prepare the transaction data in the form of *RDD[Array[String]]* by creating trajectories through the *mapCreateTrajectory* method defined above. The resulting trajectories are filtered to remove any empty ones and then reduced by key to group together trajectories pertaining to the same user. Next, they are transformed into a set of unique itemsets using the *distinct* function. This is then passed as input to the FPGrowth algorithm imported from the MLlib library, which creates a model based on the frequent itemsets and generates association rules using the *minSupport* and *minConfidence* parameters. Support and confidence are two common measures used in association rule mining to measure the strength and significance of a relationship between two items in a dataset. Support measures the frequency with which an itemset appears in the dataset, and it is calculated as the proportion of transactions in the dataset that contain the itemset. On the other hand, confidence measures the conditional probability of the consequent given the antecedent in an association rule. It is calculated as the support of the itemset containing both the antecedent and the consequent divided by the support of the antecedent alone. Finally, the method returns the frequent itemsets and the generated association rules.

```
def computeTrajectoryUsingFPGrowth(df: DataFrame,
    stringShapeMap: mutable.Map[String, String],
    minSupport: Double = 0.01, minConfidence:Double = 0.2)
    = {
  val prepareTransaction = df
    .rdd
    .map(x=>mapCreateTrajectory(x,stringShapeMap))
    .filter(x=>x._2.nonEmpty)
    .reduceByKey((x,y)=> x ++ y)

  val transactions = prepareTransaction
    .map{x=>var arr: Array[String] = Array[String]()
    x._2.foreach(x=>{
      arr = arr :+ x.poi})
    arr}
    .map(x =>  x.distinct)

  val fpg = new FPGrowth()
    .setMinSupport(minSupport)

  val model = fpg.run(transactions)

  (model.freqItemsets.collect(),model.
      generateAssociationRules(minConfidence).collect())
}
```

Listing 5.5: Computing trajectory using FPGrowth.

An excerpt of the association rules produced as output by this procedure is presented in the following, where the confidence and lift scores are reported for each rule mined from the FPGrowth algorithm.

{romanforum,stpeterbasilica} ⇒ {colosseum}: (confidence: 0.75; lift: 3.84)
{colosseum, stpeterbasilica} ⇒ {romanforum}: (confidence: 0.375; lift: 3.48)
{colosseum,stpeterbasilica} ⇒ {pantheon}: (confidence: 0.375; lift: 3.24)
{piazzadelpopolo} ⇒ {piazzadispagna}: (confidence: 0.25; lift: 2.98)

Lift is another common measure of the degree of association between two items in a dataset relative to the expected degree of association if the two items were independent of each other. A lift value greater than 1 indicates a positive association between the two items, while a lift value less than 1 indicates a negative association between the two items. The *minSupport* parameter for reproducing this output is 0.01. In the MLlib implementation of the algorithm, the minimal support level of the frequent pattern is set such that any pattern that appears more than $(minSupport * size\_of\_the\_dataset)$ times will be output.

5.3.1.2 *Hadoop implementation*

The same application described using Spark can be implemented through the MapReduce paradigm in Apache Hadoop. In particular, we define:

- a mapper, namely *DataMapperFilter*, which performs the filtering steps;
- a reducer, namely *DataReducerByDay*, which extracts the RoIs for trajectory mining;
- a main, which combines the mapper and the reducer and then applies the FPGrowth algorithm from the Apache Mahout library to mine trajectories from the RoIs extracted by the reducer.

We used the same Java implementation of the GeoUtils and KMLUtils classes described for the Spark application. For utility purposes, we defined a class for representing and dealing with Flickr data, namely *Flickr*, which provides getter methods for *user_id, longitude, latitude,* and other fields. Moreover, this class allows importing Flickr data from a string or JSON object to build a Flickr item and exporting the data as a string, as shown in Listing 5.6.

```
public class Flickr {

private final static DateTimeFormatter dateStringFormat =
    DateTimeFormat.forPattern("MMM dd, yyyy h:mm:ss a").
    withLocale(Locale.ENGLISH);

private String userId;
private double longitude;
private double latitude;
private LocalDateTime date;
private String roi;

public void importFromString(String s) {
    JSONObject json = new JSONObject(s);
    this.userId = json.getString("userId");
    this.username = json.getString("username");
    this.latitude = json.getDouble("latitude");
    this.longitude = json.getDouble("longitude");
    this.date = dateStringFormat.parseDateTime(json.
        getString("date")).toLocalDateTime();
    this.roi = json.getString("roi");
}
public void importFromFlickrJSON(String s) {
    try {
        JSONObject jsonObject = new JSONObject(s);
        this.userId = jsonObject.getJSONObject("owner").
            getString("id");
        this.username = jsonObject.getJSONObject("owner").
            getString("username");
        if (jsonObject.has("geoData")) {
            JSONObject geo = jsonObject.getJSONObject
                ("geoData");
            this.longitude = geo.getDouble("longitude");
            this.latitude = geo.getDouble("latitude");
        }
        this.date = dateStringFormat.parseDateTime
            (jsonObject.getString("dateTaken")).
            toLocalDateTime();
        this.roi = "";
```

```
    } catch (Exception e) {
    }
}

public String export() {
    JSONObject ret = new JSONObject();
    ret.put("userId", userId);
    ret.put("username", username);
    ret.put("longitude", longitude);
    ret.put("latitude", latitude);
    ret.put("date", dateStringFormat.print(date));
    ret.put("roi", roi);
    return ret.toString();
}
}
```

Listing 5.6: Flickr class.

The *DataMapperFilter* class, described in Listing 5.7, extends the *Mapper* class from the Hadoop MapReduce library and includes a few instance variables, such as a *Shape* object that represents a geospatial polygon (e.g., the geographical boundaries of the city of Rome), a *Map* object that maps location names to polygons, and *Text* objects representing output key–value pairs.

The *setup* method is called before the *map* method is executed and initializes the *romeShape* variable with a polygon based on the coordinates of Rome and the *shapeMap* variable by calling utility methods in the GeoUtils and KMLUtils classes.

The *map* method is the main processing function of the mapper. It uses the Flickr class to parse the input JSON data and extract information about the GPS coordinates of the post, the user ID, and other metadata. The method applies two filter functions, *filterIsGPSValid* and *filterIsInRome*, to check the filtering criteria, i.e., having valid GPS coordinates and being located within the vicinity of Rome. If the post passes the filters, the method uses the GeoUtils class to create a *Shape* object that represents the location of the post and checks if it is contained within one of the polygons defined in the *shapeMap* object extracted from the KML file of the PoIs in Rome. If a match is found, the PoI is set for the Flickr item. Finally, the user ID and the serialized Flickr object enriched with information about the extracted RoI are written to the output context.

```java
public class DataMapperFilter extends Mapper<LongWritable,
    Text, Text, Text> {
private Shape romeShape;
private Map<String, String> shapeMap;
private final String kmlPath = "rome.kml";
private Text outputKey = new Text();
private Text outputValue = new Text();
private final double LAT = 12.492373;
private final double LNG = 41.890251;
private final int RADIUS = 10000;


@Override
protected void setup(Mapper<LongWritable, Text, Text,
    Text>.Context context)
        throws IOException, InterruptedException {
    romeShape = GeoUtils.getCircle(GeoUtils.getPoint
        (LAT, LNG), RADIUS);
    shapeMap = KMLUtils.lookupFromKml(kmlPath);
}

public void map(LongWritable key, Text value, Context
    context)
    throws InterruptedException, IOException {
    Flickr f = new Flickr();
    f.importFromFlickrJSON(value.toString());

    if (!(filterIsGPSValid(f) && filterIsInRome(f)))
        return;

    Shape point = GeoUtils.getPoint(f.getLongitude(),
        f.getLatitude());
    boolean found = false;
    for (Entry<String, String> entry : shapeMap.entrySet())
        {
        String place = entry.getKey();
        String polygon = entry.getValue();
        try {
            Shape pol = GeoUtils.getPolygonFromString
                (polygon);
            if (GeoUtils.isContained(point, pol)) {
                f.setRoi(place);
                found = true;
            }
        } catch (IOException | ParseException e) {
```

```
            e.printStackTrace();
        }
    }

    if (found) {
        outputKey.set(f.getUserId());
        outputValue.set(f.export());
        context.write(outputKey, outputValue);
    }
}

private boolean filterIsGPSValid(Flickr f) {
    return f.getLongitude() > 0 && f.getLatitude() > 0;
}

private boolean filterIsInRome(Flickr f) {
    Point p = (Point) GeoUtils.getPoint(f.getLongitude(), f.
        getLatitude());
    return GeoUtils.isContained(p, romeShape);
}
```

Listing 5.7: DataMapperFilter class.

At this point, as described in Listing 5.8, each reducer receives for each user the set of its Flickr posts, which were previously enriched with information about the RoI. The aim of the reducer is to compute the trajectories as a sequence of RoIs. The core of the reduction phase is the *concatenateLocationsByDay* method, which builds the sorted sequence of RoIs visited on each day. The method uses a comparator to sort the Flickr items by date, and then, for each day, a string with the visited RoIs is built and added to a list, which will be returned to the *reduce* method. The *reduce* method then outputs, in the distributed context, the list of RoIs for each day separated by a blank space.

```
public class DataReducerByDay extends Reducer<Text, Text,
    NullWritable, Text> {
private Text outputValue = new Text();

@Override
public void reduce(Text key, Iterable<Text> values, Context
    context) throws java.io.IOException,
    InterruptedException {

    List<String> res = concatenateLocationsByDay(values);
    for (String s: res) {
        outputValue.set(s);
```

```java
        context.write(NullWritable.get(), outputValue);
    }
}

private static List<String> concatenateLocationsByDay
    (Iterable<Text> listItems) {
    LocalDateTime oldTimestamp = new LocalDateTime(0);
    LocalDateTime currTimestamp = null;
    List<String> ret = new LinkedList<String>();
    Set<String> s = null;
    String oldLocation = null;
    String currentLocation = null;

    List<Flickr> lf = new LinkedList<Flickr>();
    for (Text value: listItems) {
        Flickr item = new Flickr();
        item.importFromString(value.toString());
        lf.add(item);
    }

    lf.sort(new Comparator<Flickr>() {
        @Override
        public int compare(Flickr f1, Flickr f2) {
            return (f1.getDateWithoutTime().compareTo(f2.
                getDateWithoutTime())));
        }
    });
    for (Flickr f : lf) {
        currTimestamp = f.getDateWithoutTime();
        if (Days.daysBetween(oldTimestamp, currTimestamp).
            getDays() > 0) {
            if (s != null)
                ret.add(s.toString().replaceAll("\\[", "").
                    replaceAll("\\]", "").replaceAll(", ", "
                    ").trim());
            s = new HashSet<String>();
            oldLocation = null;
            oldTimestamp = currTimestamp;
        }
        currentLocation = f.getRoi();
        if (!currentLocation.equals(oldLocation)) {
            s.add(currentLocation);
            oldLocation = currentLocation;
        }
    }
```

```
if (s != null)
    ret.add(s.toString().replaceAll("\\[", "").
        replaceAll("\\]", "").replaceAll(", ", " ").trim
        ());
    return ret;
}
}
```

Listing 5.8: DataReducerByDay class.

An example of the reducer's output is shown in the following.

stpeterbasilica
piazzadispagna
mausoleumofhadrian
stpeterbasilica
capitolinehill
trevifontain
pantheon piazzadelpopolo piazzadispagna trevifontain piazzanavona
colosseum mausoleumofhadrian

Finally, the FPGrowth algorithm can be used to mine frequent trajectories. The parallel implementation of the algorithm we used is based on the work of Li *et al.* (2008), who proposed a MapReduce approach to the parallel FPGrowth (PFP) algorithm. The algorithm uses three MapReduce phases and five steps:

1. *Sharding*, which involves dividing the data into shards and distributing them on different nodes.
2. *Parallel counting*, which uses a MapReduce approach to count the support values of all items in the input dataset and stores the result in a list of frequent items sorted by frequency in descending order, called *F-List*.
3. *Grouping items*, which involves grouping the items on the F-List into groups based on their support values.
4. *Parallel FPGrowth*, which is the key step of PFP, where a MapReduce pass is used to perform parallel FPGrowth on group-dependent shards.
5. *Aggregating*, which involves aggregating the results generated in the previous step to produce the final result.

The *main* method in Listing 5.9 finally combines the mapper, the reducer, and the FPGrowth algorithm.

```
public class Main extends Configured implements Tool {

private static String inputPath = "FlickrRome.json";
private static String trajOutputPath = "outputMR/";
private static String fpOutputPath = "fpOutput/";


@Override
public int run(String[] args) throws Exception {

    int minimumSupport = Integer.parseInt(args[0]);
    int maxPatterns = Integer.parseInt(args[1]);
    Configuration conf = new Configuration();
    conf.set("fs.defaultFS", "file:///");

    /** Trajectory Job configuration */
    Job traJob = Job.getInstance(conf, "TrajectoryJob");

    /* FileSystem setting */
    FileInputFormat.addInputPaths(traJob, inputPath);
    FileOutputFormat.setOutputPath(traJob, new Path
        (trajOutputPath));
    FileSystem fs = FileSystem.get(conf);
    if (fs.exists(new Path(trajOutputPath)))
        fs.delete(new Path(trajOutputPath), true);
    traJob.setMapperClass(DataMapperFilter.class);
    // Set the output class of key and value for the mapper
    traJob.setMapOutputKeyClass(Text.class);
    traJob.setMapOutputValueClass(Text.class);

    traJob.setReducerClass(DataReducerByDay.class);
    // Set the output class of key and value for the reducer
    traJob.setOutputKeyClass(Text.class);
    traJob.setOutputValueClass(Text.class);

    boolean completed;
    completed = traJob.waitForCompletion(true);
    if (!completed)
        return 1;

    /** Run FPGrowth algorithm */
    if (fs.exists(new Path(fpOutputPath)))
        fs.delete(new Path(fpOutputPath), true);
```

```
FpGrowth.runFpGrowth(trajOutputPath, fpOutputPath,
    minimumSupport, maxPatterns);
    return 1;
}

public static void main(String[] args) throws Exception {
    int res = ToolRunner.run(new Configuration(),
        new Main(), args);
    System.exit(res);
}
}
```

Listing 5.9: Main class.

By setting the *minSupport* parameter equal to 3 and a high value for the *maxPatterns* parameter (e.g., 500), this Hadoop implementation generates the same rules as the Spark one.

5.3.2 Streaming application: Apache Storm vs. Apache Spark Streaming

The proposed application implements a network intrusion detection system (NIDS) aimed at early detection of network intrusion and malicious activities. Network intrusion detection is a crucial part of network management to improve security and ensure the quality of service. Typically, these systems use data mining or machine learning techniques to automatically detect attacks against computer networks and systems.

5.3.2.1 Storm implementation

A Storm application requires defining three entities: spouts, bolts, and the topology. The proposed application is implemented by using a Storm topology that is composed of one spout and two bolts:

- *ConnectionSpout*, which is the data source in the topology. This spout streams connections that come from a firewall or are stored in a log file, and each record is forwarded as a tuple to the next bolt. In this example, a connection is described by a set of 41 features, which describe the characteristics of a network connection (e.g., duration, protocol type, and service).
- *DataPreprocessingBolt*, which receives the tuples from the spout and performs some preprocessing steps on the data. Specifically,

the preprocessing phase includes the conversion of categorical features into numerical ones and the standardization of these features for the machine learning model.

- *ModelBolt*, which performs the classification through a random forest model that is trained offline and used in real time for monitoring connections to detect potential malicious activities.

As stated, the training phase is performed offline using the Python *scikit-learn* library since Storm does not provide any native machine learning libraries. All the trained models (i.e., the standard scaler for numerical features, the label encoder for categorical features, and the random forest model) are dumped in files using the Python *pickle* module. Hence, the Storm multi-language protocol can be adopted to integrate the trained models into a topology implemented in a JVM language. Figure 5.2 shows the whole architecture of the proposed application.

The Storm topology, implemented using Java, is shown in Listing 5.10. The code starts by creating a configuration object that can be used to specify various parameters for the topology, such as the number of worker nodes in the cluster. The next step is to create the topology itself using the *TopologyBuilder* class. The topology is represented as a DAG of the compute components, with data flowing from spouts to bolts. The code then adds the three components defined above, namely a spout and

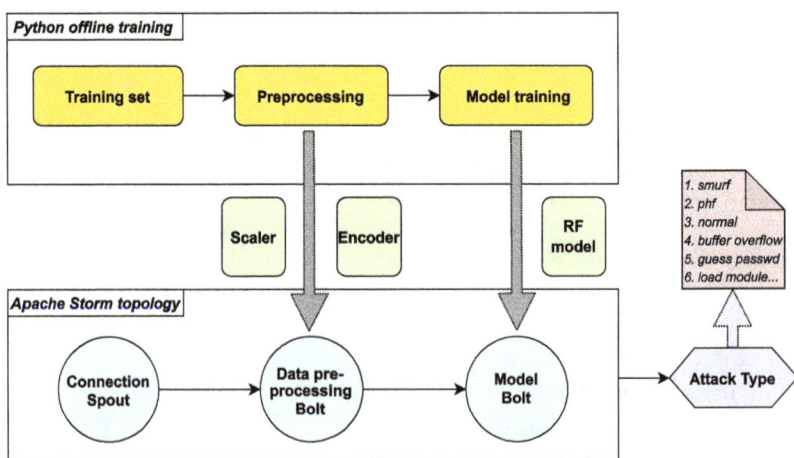

Fig. 5.2.   Architecture of the proposed Storm application.

two bolts, to the topology. The *shuffleGrouping* method between *ConnectionSpout*, *DataPreprocessingBolt*, and *ModelBolt* specifies that the data should be randomly shuffled before being sent to the next component, allowing for a more even data distribution. Once the topology is fully defined, it is submitted to the cluster using the *StormSubmitter.submitTopology* method, which takes three arguments, i.e., the topology name, the configuration object, and the topology itself, created using the *createTopology()* method. The topology, once running on a cluster, will process the data as they become available.

```
public class IntrusionTopology {
    public static void main(String[] args) {
        // Build and submit the topology to a cluster
        Config conf = new Config();
        conf.setNumWorkers(20);
        TopologyBuilder builder = new TopologyBuilder();
        builder.setSpout("spout", new ConnectionSpout());
        builder.setBolt("process", new DataPreprocessingBolt
            ())
                .shuffleGrouping("spout");
        builder.setBolt("model", new ModelBolt())
                .shuffleGrouping("process");
        StormSubmitter.submitTopology("IntrusionDetection",
            conf, builder.createTopology());
    }
}
```

Listing 5.10: Storm topology.

Each spout node must specify the collector used to emit the tuples (the method *open*), how to emit the next tuple (the method *nextTuple*), and declare the output fields for the tuples it emits (the method *declareOutputFields*), as shown in Listing 5.11. A tuple emitted by the spout represents the set of 41 features that describe a connection (e.g., duration, protocol type). Each field of the tuple is assigned a name previously stored in an array of strings, namely *field_names*.

```
public class ConnectionSpout implements IRichSpout {
    private SpoutOutputCollector collector;
    // Define the name of each column in the training data
    private String[] field_names = new String[]{"duration",
        "protocol_type", "service", ...};
```

```
@Override
public void open(Map conf, TopologyContext context,
    SpoutOutputCollector collector) {
    // Define the collector to be used for emitting
        tuples
    this.collector = collector;
}

@Override
public void nextTuple() {
    // Read from a log file
    while ((str = reader.readLine()) != null) {
        String[] fields = str.split(",");
        // Emit a tuple from input file
        this.collector.emit(new Values(fields[0], ... ,
            fields[40]));
    }
}

@Override
public void declareOutputFields(OutputFieldsDeclarer
    dec) {
    // Declare the name of each field of the tuples
    dec.declare(new Fields(field_names[0], ...,
        field_names[40]));
}
}
```

Listing 5.11: Storm *ConnectionSpout*.

Each tuple is emitted by the spout and will be processed by
the subsequent bolts as declared in the topology. In this case, the
class *DataPreprocessingBolt* in Listing 5.12 is a proxy for the Python
bolt defined in Listing 5.13. The Python script of the bolt processes a tuple by applying the transformations of a set of models loaded from the disk (e.g., the encoders for categorical features
and the scalers for numerical features), which have been previously
trained offline on a training dataset of normal and malicious connections. The multi-language protocol only requires the bolt to specify
the script to be executed by invoking the *super* method (see Listing 5.12), while all the application logic is contained in the Python
script.

```
public class DataPreprocessingBolt extends ShellBolt
    implements IRichBolt {
    public DataPreprocessingBolt() {
        // Use the Multi-Language protocol to run a Python
            script
        super("python3", "preprocessingBolt.py");
    }
}
```

Listing 5.12: Storm *DataPreprocessingBolt*.

```
class preprocessingBolt(storm.BasicBolt):
    # Load encoders for protocol type, service, and flag
    label_enc_prot = pickle.load(open("label_enc_prot",
        'rb'))
    label_enc_serv = pickle.load(open("label_enc_serv",
        'rb'))
    label_enc_flag = pickle.load(open("label_enc_flag",
        'rb'))

    # Load the standard scaler for numerical features
    standard_scaler = pickle.load(open("standard_scaler",
        'rb'))
    # Mark nominal, binary, and numerical features
    nominal_idx, binary_idx = [1, 2, 3], [6, 11, 13, 14, 20,
        21]
    numerical_idx = list(set(range(41)).difference
        (nominal_idx).difference(binary_idx))

    def process(self, tuple):
        # Encode categorical features
        # protocol type = nominal_idx[0]
        prot_type = label_enc_prot.transform([tuple.values
            [nominal_idx[0]]])[0]
        # service = nominal_idx[1]
        serv = label_enc_serv.transform([tuple.values
            [nominal_idx[1]]])[0]
        # flag = nominal_idx[2]
        flag = label_enc_flag.transform([tuple.values
            [nominal_idx[2]]])[0]

        # Scale numerical features
        scaled_features = standard_scaler.transform
            (np.reshape([float(tuple.values[i]) for i in
                numerical_idx], (1, -1)))[0]
```

```
        # Emit the tuple after processing
        storm.emit([scaled_features[0], prot_type, serv,
            flag, ... ])

preprocessingBolt().run()
```

Listing 5.13: Python *DataPreprocessingBolt.*

Finally, the ModelBolt in Listing 5.14 acts similarly to the DataPreprocessingBolt defined in Listing 5.12. In particular, it applies the random forest model trained offline by using the Python script described in Listing 5.15. The predicted connection type from a set of 23 types, including smurf, buffer overflow, and guess password, enables the network security infrastructure to react and mitigate possible threats.

```
public class ModelBolt extends ShellBolt implements
    IRichBolt {
    public ModelBolt() {
        // Use the Multi-Language protocol to run a Python
            script
        super("python3", "modelBolt.py");
    }
}
```

Listing 5.14: Storm *ModelBolt.*

```
class modelBolt(storm.BasicBolt):
    # Load the RF model from disk
    model = pickle.load(open("rf_model", 'rb'))

    def process(self, tuple):
        # Predict the connection type from a set of 23 types
        prediction = model.predict(np.reshape([tuple.
            values[0], tuple.values[1],..., tuple.values
                [40]], (1, -1)))[0]

        # Emit the predicted connection type
        storm.emit([prediction])

modelBolt().run()
```

Listing 5.15: Python *ModelBolt.*

5.3.2.2 *Spark Streaming implementation*

Following our discussion on implementing an intrusion detection system using Storm, we now show how the same real-time application can be coded using Spark Streaming. Spark Streaming is a library provided in Apache Spark for scalable, high-throughput, and fault-tolerant stream processing. It supports multiple data ingestion sources, including Apache Kafka, Amazon Kinesis, and TCP sockets, and leverages the power of the distributed processing capabilities of Apache Spark and its built-in libraries, such as Spark SQL and Spark MLlib, to provide an end-to-end stream processing solution. Once ingested, the data can be transformed and processed using advanced algorithms (e.g., machine learning and graph processing) to produce meaningful insights and predictions. The processed data can be stored in various storage systems, such as Apache Cassandra, Apache HBase, or Amazon S3, for further analysis and decision-making. Spark Streaming provides a high-level abstraction for continuous data streams, called Discretized Stream (DStream). A DStream is a continuous sequence of RDDs, where each RDD collects data from a given time window, and all operations performed on a DStream are forwarded to the underlying RDDs. This abstraction allows developers to perform transformations and actions on incoming data in real time, without the need to persist all the data in memory.

The first part of the application comprises training an offline machine learning model, specifically a random forest classifier, on the training data. In particular, the following steps are performed:

(1) Load the CSV file into a DataFrame using Spark's CSV data source and a schema for assigning a name to each column.
(2) Randomly split the data into training and testing datasets to train and further evaluate the model before using it in a real-time setting.
(3) Define the preprocessing steps:
   - Use a *VectorAssembler* to combine multiple columns into a single column of features.
   - Use a *StringIndexer* to convert the target categorical label column into a numeric label column.
   - Use a *StandardScaler* to scale the features to have a zero mean and unit variance.

(4) Define *RandomForestClassifier* as the machine learning model to be trained.

(5) Create a pipeline to combine the preprocessing steps in order and train the model on the training data.

(6) Use the trained pipeline to make predictions on the testing data.

(7) Evaluate the model's accuracy on the testing data using *MulticlassClassificationEvaluator*.

(8) Save the trained pipeline and the trained *RandomForestClassifier* model to disk.

Listing 5.16 shows how the preprocessing pipeline and the random forest classifier can be trained, evaluated, and saved to disk for further reuse using Spark MLlib.

```scala
val df = spark.read.format("csv").option("header",
    "false").schema(connectionSchema).load(path_df)
val Array(trainingData, testData) = df.randomSplit(Array
    (0.7, 0.3))

// Define the preprocessing steps
val assembler = new VectorAssembler()
  .setInputCols(Array("duration", "protocol_type",
      "service", "..."))
  .setOutputCol("features")
val indexer = new StringIndexer().setInputCol
    ("attack_type").setOutputCol("indexed_label")
val scaler = new StandardScaler().setInputCol
    ("features").setOutputCol("scaledFeatures").
        setWithStd(true).setWithMean(false)
val rf = new RandomForestClassifier().setFeaturesCol
    ("scaledFeatures").setLabelCol("indexed_label")

// Put the preprocessing steps into a pipeline
val pipeline = new Pipeline().setStages(Array(assembler,
    indexer, scaler, rf))

// Train the pipeline on the training data
val pipelineModel = pipeline.fit(trainingData)
val predictions = pipelineModel.transform(testData)
// Save the pipeline and the Random Forest model to the
    disk for reuse
pipelineModel.write.overwrite().save("pipeline_model")
pipelineModel.stages(2).asInstanceOf
    [RandomForestClassifier].write.overwrite().save
        ("rf_model")
```

```
// Evaluate the accuracy on test data
val evaluator = new MulticlassClassificationEvaluator()
  .setLabelCol("indexed_label")
  .setPredictionCol("prediction")
  .setMetricName("accuracy")
val accuracy = evaluator.evaluate(predictions)

spark.stop()
```

Listing 5.16: Offline training of a random forest classifier for intrusion detection using Spark MLlib.

After training and testing the machine learning model, it can now be used for the real-time detection of malicious connections. To do this, the model that was previously saved to disk has to be loaded and used in conjunction with the Spark Streaming APIs, which are used to handle the input stream. Listing 5.17 shows the real-time part of the Spark Streaming application.

```
val conf = new SparkConf().setMaster("...")
// StreamingContext with a batch interval of 1 second
val ssc = new StreamingContext(conf, Seconds(1))
// Create a DStream that will connect to hostname:port
val data: ReceiverInputDStream[String] = ssc.
    socketTextStream(hostname, port)
// Process data in window of windowLength and slideInterval
val dataInWindow = data.window(Seconds(30), Seconds(10)).map
    {line =>
        val col = line.split(",")
        ConnectionTest(col(0).toDouble, col(1).toInt, ...,
            col(40).toInt) }
// Load trained model and scaler
val pipelineLoaded = PipelineModel.load("pipeline_model")
// Apply transformations on each underlying RDD and get
    predictions
dataInWindow.foreachRDD { rdd =>
    pipelineLoaded.transform(rdd.toDF()).show()  }
}

// Start the computation and wait for it to terminate
ssc.start()
ssc.awaitTermination()
```

Listing 5.17: Network intrusion detection using Spark Streaming.

The first step in the code is to create a *SparkConf* object and set the master node for the Spark cluster. Then, a *StreamingContext* object is created with a batch interval of 1 s, which is the entry point for all streaming operations. Once a context has been defined, the subsequent steps involve the following operations: (i) creating input DStreams to specify the input sources; (ii) applying transformation and output operations to DStreams to define the streaming computations; (iii) starting the data reception and processing by invoking the *start()* operation on the streaming context; and (iv) waiting for the processing to stop (either manually or due to an error) via the *awaitTermination()* operation. Additionally, the processing can be manually stopped by using *streamingContext.stop()*.

StreamingContext is used to ingest and collect incoming data from a specified host and port via a socket connection in the form of a *ReceiverInputDStream* of strings. Data can also be obtained from files, but the filesystem has to be compatible with the HDFS API (e.g., HDFS, S3, and NFS). To create a DStream in this scenario, the *streamingContext.textFileStream(dataDirectory)* command can be used. This allows Spark Streaming to monitor the specified directory and process any files created in that directory. All files located directly under this path will be processed as they are detected, but updates to files within the current window will be ignored. It is important to keep in mind that the more files in a directory, the longer the scanning time for changes will be, even if none of the files have been modified. Lastly, it is worth noting that only the files whose modification time falls within the current window will be included in the stream.

The received data are then processed to extract relevant features and organized into windows of length 30 s with a sliding interval of 10 s. Each line of incoming data is split into columns using the *split* method and then parsed into a *ConnectionTest* object, which represents a connection with all the fields previously described. The trained pipeline is loaded from the disk and then applied to each RDD in the *DStream*. The *transform* method of the scaler is used to normalize the features, and the *predict* method of the *RandomForestClassifier* model is used to make predictions on the processed data. Another important consideration is that the computation of a stream is triggered only when an output operator is called; otherwise, without an output operator on DStream, no computation is invoked, similar to actions for RDDs. Currently, the output operations that can be invoked are *print*, *saveAsTextFiles*, *saveAsObjectFiles*, *saveAsHadoopFiles*, and *foreachRDD*. The *foreachRDD* is the most generic output operator and is used to apply a function to every RDD produced by the stream. The function passed as input to the operator is responsible for sending the data from each RDD to an external system, such as saving it to files or transmitting it over the network to a database.

One of the features offered by Spark Streaming is the possibility to perform windowed operations, similar to those provided by Apache Storm. Windowed operations allow processing a sliding window of data, combining all RDDs that fit within the window and processing them to produce a windowed DStream. To perform a windowed operation, programmers must specify the duration of the window and the interval in which the windowed operation is performed. One of Spark Streaming's main advantages over Apache Storm is its full integration with MLlib, which facilitates the use of a wide range of algorithms for offline learning without the need for external libraries. Additionally, Spark Streaming natively supports streaming machine learning algorithms that can simultaneously learn and predict based on a stream of data (e.g., streaming linear regression and streaming K-Means). The library also supports the Scala programming language, which leads to more compact and readable code. However, it is worth noting that Spark Streaming adopts micro-batch processing, as opposed to pure streaming processing, such as in Apache Storm. For this reason, the batch interval is one of the critical hyperparameters to set in a Spark Streaming application. To ensure the stability of the application, it should process data as fast as they are received, meaning that batches should be processed at the same rate they are generated. Due to this, data rates impact the batch interval to be used.

5.3.3 *SQL application: Apache Hive vs. Apache Spark SQL*

One of the leading trends in social media research is the analysis of geotagged data to determine whether users have visited or not interesting locations, commonly called PoIs, such as tourist attractions, shopping malls, squares, and parks. As introduced in Section 5.3.1, to match user trajectories with PoIs, it is often useful to define the so-called RoI, which represents the boundaries of a PoI's area.

Here, we discuss a data analytics application for extracting a suitable RoI for a PoI by analyzing a large set of geotagged posts gathered from social media. A post g is geotagged if it contains a pair of coordinates (latitude and longitude) identifying the place from where g was created. Moreover, a post g can be associated with a PoI p if its text or tags refer to p. As an example, in their posts, social network users identify the *Colosseum* with different keywords, such as *Coliseum*, *Coliseo*, or *Colise*, and synonymous, such as *Flavian Amphitheatre* or *Amphitheatrum Flavium*.

Once all the coordinates of social media posts that refer to a specific PoI are extracted, the RoI is obtained by using a spatial clustering algorithm (e.g., DBSCAN). The RoI is the polygon enclosing the largest cluster of points discovered by the algorithm.

5.3.3.1   *Hive implementation*

As discussed in Section 4.6.1, Hive enables incorporating advanced data processing in the query by defining different types of functions (i.e., UDF, UDAF, or UDTF). In particular, developers can define two types of UDAFs: simple and generic. Simple UDAFs are rather simple to write but can incur performance issues due to the use of Java Reflection and do not allow features such as variable-length argument lists. Instead, generic UDAFs allow all these features but are perhaps not quite as intuitive to write as simple UDAFs.

To implement the application natively in Hive, we created two custom functions:

- *GeoDataUDF*, which, given the text of a post, checks if it contains at least one tag referring to the PoI.
- *DbscanUDAF*, which, given the set of valid coordinates extracted from the posts, exploits the DBSCAN algorithm for extracting the cluster representing the RoI of the PoI.

GeoDataUDF is shown in Listing 5.18. In particular, the UDF function is written by extending the abstract class *GenericUDF*, which, unlike the simplest class UDF, uses the concept of *ObjectInspector*. An instance of ObjectInspector for a specific type, such as a native Java object (e.g., String, Double, and Collection), allows programmers to directly access the internal fields and methods of that type. Therefore, all interactions with the data passed into UDFs are performed through ObjectInspectors, which allow reading input values from an UDF parameter and writing output values. Looking at the code in Listing 5.18 in detail, it should be noted that we override the following methods from the GenericUDF class:

- *initialize*: It initializes the GenericUDF object. In particular, this method is called once and only once per GenericUDF instance. It is responsible for checking input arguments (types and length) and returning an ObjectInspector for the return value.
- *evaluate*: It implements the logic of the GenericUDF function. When invoked, it applies the function with the arguments to input objects, which can be inspected by the ObjectInspector passed in the *initialize* method.
- *getDisplayString*: It defines the string to be displayed when invoking the *EXPLAIN* command in Hive. The Hive EXPLAIN command is used to display the execution plan that the Hive query engine generates and uses while executing any query. Such a command is then very useful to understand the query execution flow when you are trying to optimize it.

```
public class GeoDataUDF extends GenericUDF {

    private StringObjectInspector input;
    private StringObjectInspector input2;

    @Override
    public ObjectInspector initialize(ObjectInspector[]
        args) throws UDFArgumentException {
        // check to make sure the input has 1 argument
        if (args.length != 2)
            throw new UDFArgumentException("input must have
                length 2");
        // set ObjectInspector's objects from the input
            arguments, checking before they are both strings
        if (!(args[0] instanceof StringObjectInspector))
            throw new UDFArgumentException("input 1 must be
                a string object");
        if (!(args[1] instanceof StringObjectInspector))
            throw new UDFArgumentException("input 2 must be
                a string object");
        this.input = (StringObjectInspector) input;
        this.input2 = (StringObjectInspector) input2;
```

```java
        // return the ObjectInspector to be used as the
            return value
        return PrimitiveObjectInspectorFactory.
            javaStringObjectInspector;
    }

    @Override
    public Object evaluate(DeferredObject[] args) throws
        HiveException {
        // Check if number and value of the input arguments
            are correct
        if (input == null || input2 == null || args.length
            != 2 || args[0].get() == null || args[1].get()
            == null)
            return null;
        // Get the textual content of a social media post
            from args[0]
        String postText = input.getPrimitiveJavaObject(args
            [0].get()).toString().toLowerCase();
        // Get the list of keywords identifying a PoI from
            args[1]
        String[] forwardKeywordsArray = input2.
            getPrimitiveJavaObject(args[1].get()).toString()
            .split(",");
        // Check if at least one keyword is contained in the
            post's text
        for (String k : forwardKeywordsArray) {
            if (postText.contains(k.toLowerCase())) {
                // return the first keyword identifying the
                    PoI
                return forwardKeywordsArray[0];
            }
        }
        // Return null if the post can't be associated to
            the PoI
        return null;
    }

    @Override
    public String getDisplayString(String[] strings) {
        return getStandardDisplayString("geoData", strings);
    }
}
```

Listing 5.18: GeoDataUDF function for processing social media posts.

Listing 5.19 shows the implementation of the *DbscanUDAF* function, which has been implemented as a simple UDAF. The class *DbscanUDAF* defines the static class *DbscanUDAFEvaluator* within it, which implements the *UDAFEvaluator* interface. Hive executes the UDAF function by invoking in sequence the following methods of the evaluator class:

- *init*, which initializes the UDAF evaluator by setting an empty partial result (i.e., the state of the partial aggregation), indicating that no values have been aggregated yet;
- *iterate*, which is invoked every time there is a new value to be aggregated;
- *terminatePartial*, which is called to return the partially aggregated result;
- *merge*, which is invoked when Hive has to combine one partial aggregation with another;
- *terminate*, which is called when the final result of the aggregation must be computed.

Specifically, during the *iterate* method, the function aggregates the available values by storing the coordinates (latitudes, longitudes) and parameters (*eps* and *minPts*) inside a static object (*DbscanGeo-Dataset*, which is used to store the state of the partial aggregation). Then, Hive aggregates other local aggregations by invoking the *merge* method. Finally, the *terminate* method is called to produce the final result by executing an instance of DBSCAN. In particular, we used an implementation of DBSCAN for geospatial clustering, which is included in the ELKI framework[2].

```
@Description(name = "dbscan",
        value = "_FUNC_(lat, lng, eps, minPts) - Returns the
            largest cluster found by dbscan in a list of
            geo-points"
)
public class DbscanUDAF extends UDAF {

    public static class DbscanUDAFEvaluator implements
        UDAFEvaluator {
```

---

[2]https://elki-project.github.io/.

```
// static class used to store the state of
    aggregation
public static class DbscanGeoDataset {
    private List<String> points = new
        LinkedList<>();
    private double eps = 0;
    private int minPts = 0;
}

// static object that stores the state of local
    aggregation
private DbscanGeoDataset dataset = null;

// Constructor of the evaluator
public DbscanUDAFEvaluator() {
    super(); init();
}

// Initialize the state of the aggregation as empty
    (no values have been aggregated yet)
@Override
public void init() {
    dataset = new DbscanGeoDataset();
}

// Invoked to aggregate each new coordinate. The
    DBSCAN's parameters are maintained inside the
    aggregation state.
public boolean iterate(double latitude, double
    longitude,
double eps, int minPts) throws HiveException {
    if (dataset == null)
        throw new HiveException("DBSCAN dataset is
            not initialized");
    if (latitude > 0 && longitude > 0) {
        // Add valid coordinates and parameters to
            the local state
        dataset.points.add(longitude + "," +
            latitude);
        dataset.eps = eps;
        dataset.minPts = minPts;
        return true;
    } else
        return false;
}
```

```
// Return the partially aggregated results
public DbscanGeoDataset terminatePartial() {
    return dataset;
}

// Merge two partial results (i.e., merge points
   that have been found in a single dataset)
public boolean merge(DbscanGeoDataset other) throws
    HiveException {
    if (other != null) {
        this.dataset.points.addAll(other.points);
        this.dataset.minPts = other.minPts;
        this.dataset.eps = other.eps;
    }
    return true;
}

// Calculate the final result by executing an
   instance of DBSCAN
public String terminate() throws HiveException {

    // Prepare an instance of DBSCAN using the given
       parameters and the fully-aggregated list of
       points
    ElkiDBSCAN dbscan = new ElkiDBSCAN(
            this.dataset.points.stream().map(p -> {
                String[] data = p.split(",");
                return new GeoPoint(Double.
                    parseDouble(data[0]),
                        Double.parseDouble(data[1]));
            }).collect(Collectors.toList()),
            this.dataset.eps, this.dataset.minPts);

    // Run the DBSCAN algorithm
    dbscan.cluster();

    // Find the largest cluster
    List<GeoCluster> clusters = dbscan.
        getAllGeoClusters(false);
    int max = -1;
    GeoCluster maxCluster = null;
    for (GeoCluster c : clusters) {
        if (c.getPoints().size() > max) {
            maxCluster = c;
```

```
                    max = c.getPoints().size();
            }
        }
        // Convert the cluster into a polygon in KML
            format
        if (maxCluster != null) {
            try {
                Shape clusterShape = GeoUtils.convexHull
                    (maxCluster.getPoints());
                return KMLUtils.serialize(clusterShape);
            } catch (IOException e) {
                return "ERROR_NOT_FOUND";
            }
        } else return "NOT_FOUND";
    }
}
}
```

Listing 5.19: GeoDataUDF function for processing social media posts.

5.3.3.2 *Spark SQL implementation*

Spark SQL is a module for structured data management and processing. It differs from the RDD API in that it extends Spark with optimizations based on information about the structure of data and the computations performed. Developers can interact with Spark SQL via SQL statements and the Dataset API using the same execution engine. In addition to SQL query execution, Spark SQL can also be used to read data from an existing Hive environment. Spark SQL provides two high-level abstractions, namely *Dataset* and *DataFrame*, as detailed in Section 4.3.1.1. The data types are automatically inferred, but developers can provide an explicit schema via a *StructType*, matching the structure of the DataFrame in named columns. Spark SQL supports the vast majority of Hive features, such as UDFs and DataFrame operations. Listing 5.20 shows how the RoI mining application discussed in the previous section can be implemented with the Spark SQL Dataset API using the Python programming language.

```
import findspark
findspark.init('/opt/spark')
from pyspark.sql import SparkSession

# Initialize a Spark session with Hive support
spark = SparkSession.builder\
    .enableHiveSupport()\
    .getOrCreate()

# Load Flickr posts from HDFS
flickrData=spark.read.json("hdfs://master/user/custom/
    allFlickrRome.json")

# Create Hive Internal table
flickrData.write.mode("overwrite").saveAsTable("posts")

# Create UDF and UDAF functions from a JAR file stored on
    HDFS
spark.sql("ADD JAR hdfs://master/user/custom/HiveUDF-1.0-
    SNAPSHOT-jar-with-dependencies.jar")
spark.sql("CREATE FUNCTION GeoData AS 'it.unical.dimes.
    scalab.hive.udf.GeoDataUDF'");
spark.sql("CREATE FUNCTION GeoDbscan AS 'it.unical.dimes.
    scalab.hive.udf.DbscanUDAF'");

# Define a list of keywords used by social media users to
    identify the Colosseum
keywords = "colosseum, colosseo, colis, collis,
    amphitheatrum flavium, colasaem, coliseo, coliseu,
    coliseum, coliseus, ..."

# Find a RoI for the Colosseum
spark.sql("SELECT GeoDbscan(longitude, latitude, 150, 50) AS
    cluster FROM posts WHERE \
        GeoData(description,'"+keywords+"') IS NOT NULL
            AND latitude > 0 AND longitude > 0").show
                (truncate=False)
```

Listing 5.20: RoI mining application in Hive with Spark SQL.

Figure 5.3 shows an example of RoI found by applying the Hive code above to a dataset of geotagged posts gathered from Flickr.

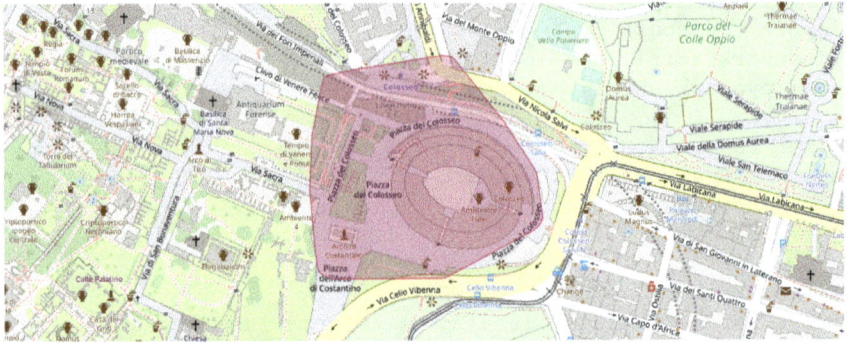

Fig. 5.3. Example of RoI discovered for the Colosseum using the *DbscanUDAF* function on a set of social media posts from Flickr.

Source: Image from OpenStreetMap, https://openstreetmap.org/copyright.

It is worth noting that Spark SQL always uses the same engine when executing a query, regardless of which API is adopted (i.e., SQL or Dataset API). This kind of uniformity brings a major benefit to developers, as they can easily switch between the two APIs depending on which one is most suitable for expressing a certain transformation. The ease with which Spark makes it possible to write data processing functions also suggests that it is much more practical to use than Hive. In fact, using UDF/UDAF in Hive for complex data analytics functions can be very difficult to write compared to Spark. This highlights the true nature of Hive, which is a data warehouse system for reading, writing, and managing up to petabytes of structured data but not designed for complex data mining or data analytics operations. Although Hive provides a large set of data aggregation and transformation functions that can be integrated directly into SQL queries, they do not cover cases where it is necessary to perform complex data processing and mining algorithms: In these cases, it is much more convenient to use Spark with Spark SQL.

5.3.4 *Graph application: MPI vs. Apache Spark GraphX*

In this section, we discuss the implementation of an application of extractive summarization, comparing two programming tools effectively exploitable for efficient graph computation, namely MPI and the GraphX library available within the Spark framework.

Extractive summarization is a type of text summarization technique that involves identifying and extracting the most important information from a document and presenting it in a concise summary. Specifically, the summary is created by arranging the most relevant sentences or phrases from the original text. These summarization techniques are generally more lightweight with respect to those that generate new text, i.e., abstractive summarization techniques, which involve understanding the meaning and context of the original text and creating a summary as written by humans through more advanced natural language processing techniques, such as large self-supervised language models. Extractive summarization has a number of relevant applications. To name a few, it is used in news article summarization, where a large amount of text must be quickly and accurately summarized for readers, and also in academic research to help researchers quickly understand the key findings and contributions of a paper.

In this section, we focus on TextRank (Mihalcea and Tarau, 2004), a summarization technique that represents the text as a graph of semantically connected sentences and extracts a summary by identifying the top-k most representative sentences by PageRank. Formally, given a text $\mathcal{T}$ to be summarized, the algorithm creates a graph $G = <S, E>$, where $S = s_1, \ldots, s_n$ are the n sentences present in $\mathcal{T}$. The set E, containing the edges of the graph, is created by connecting each pair of sentences in S, where each connection is associated with a weight $w_{i,j}$ representing the textual similarity between s_i and s_j. The similarity between two sentences s_i and s_j is given by the following formula:

$$w_{i,j} = \frac{|s_i \cap s_j|}{\log(|s_i + 1|) + \log(|s_j + 1|) + 1}$$

According to the formula, the similarity between two sentences is directly proportional to the words they have in common (i.e., their intersection) but is inversely proportional to their length, following the assumption that very long phrases may have a high intersection with each other, which does not necessarily entail a high similarity. So, the edge set will be the set of triples $E = <s_i, s_j, w_{i,j}> \forall s_i, s_j \in S \times S$, leading to an undirected, complete, weighted graph of semantically connected sentences.

Given this graph, the PageRank algorithm (Brin and Page, 1998) (discussed in detail in Section 4.5.1.4) can be leveraged to extract the most important vertices, i.e., the most representative sentences of the original text. Finally, these sentences are concatenated and returned in output as the extracted summary. In the following, we compare two different implementations of the described algorithm, which were developed using the MPI framework and the Pregel APIs of the GraphX library. The implementations are explained in detail, along with a comparison focused on the main advantages and drawbacks arising from the use of both development tools.

5.3.4.1   *GraphX implementation*

We first discuss the implementation of the application in Apache Spark using the Scala language and the GraphX library for parallel graph computation based on the bulk synchronous parallel (BSP) message-passing abstraction. We used the Pregel-like APIs provided by GraphX to implement a weighted version of the PageRank algorithm, in which the strength of the connection between two nodes is considered in the computation of the score for each node. The main parts making up the developed implementation are described as follows.

First, we created the Spark context and defined a simple Scala method to measure the textual similarity between two given sentences obtained by the formula introduced above (see Listing 5.21). We basically divide the cardinality of the intersection by the sum of the log lengths of the two input phrases. The denominator is adjusted to avoid log zero or division by zero.

```
//Create Spark session
val spark = SparkSession
    .builder.master("local")
    .appName("Spark-GraphX-TextRank")
    .getOrCreate()
val sc: SparkContext = spark.sparkContext

// define sentence similarity score
def sentenceSimilarity(s1: String, s2: String): Double = {
```

```
val words1 = s1.split(" [\\s,.;:?!]+").map(_.toLowerCase)
    .toSet
val words2 = s2.split(" [\\s,.;:?!]+").map(_.toLowerCase)
    .toSet
val commonWords = words1.intersect(words2).size
val den = log(words1.size + 1) + log(words2.size + 1)
    + 1
commonWords.toDouble / den
}
```

Listing 5.21: Define Spark context and sentence similarity.

Afterward, as shown in Listing 5.22, we create the fully connected graph of sentences from the input file. This graph is created
by reading the input text file and extracting the different sentences
it contains (i.e., the set of vertices of the sentence graph). Then,
each sentence is associated with its index, and all possible pairs of
sentences are determined as the Cartesian self-product of the set of
extracted sentences. Each obtained pair will form an edge of the fully
connected graph of sentences. Finally, the weight of each edge is computed using the method provided in Listing 5.21, and the graph is
returned as output.

```
def buildGraph(input_path: String): Graph[String, Double] =
    {
    // get sentence from textual file to be summarized
    val input_sentences = sc.textFile(input_path)
    .flatMap(line =>line.split('.'))
    // each sentence is a vertex
    val vertices = input_sentences.zipWithIndex.map {
        case (sentence, index) => (index, sentence)}
    // each pair of sentences is an edge weighted with
        pairwise similarity
    val pairs = vertices.cartesian(vertices)
    .filter { case ((i1, _), (i2, _)) => i1 != i2 }
    .map { case ((i1, s1), (i2, s2)) => (i1, i2,
        sentenceSimilarity(s1, s2)) }
    .map { case (i1, i2, sim) => Edge(i1, i2, sim) }
    Graph(vertices, pairs)
}
```

Listing 5.22: Building the sentence graph starting with an input text file.

Once created, the sentence graph is prepared for PageRank computation, as shown in Listing 5.23. In particular, the edge weights are normalized such that, for each node, the sum of the weights of the outgoing edges is equal to 1. This is achieved by creating a map containing, for each node, the sum of the weights of its out-edges, obtained by leveraging the *aggregateMessages* method provided by GraphX. In this case, *aggregateMessages* allows for processing each edge of the graph, sending the edge weight to its source node, which aggregates the messages received by all its outgoing edges using a sum as the reduce function: *(a,b)* $=>$ *a+b*. Afterward, a *map* function is used to normalize each edge based on its source node and the corresponding value stored in the map. Finally, each node is assigned the initial guess through a *mapVertices* operation.

```scala
// normalize edge weights and set initial guess
def prepareGraph(graph: Graph[String, Double]):
    Graph[Double, Double] = {
    // compute a map of normalization factors
    val outgoingEdgesSum = graph.aggregateMessages[Double](
        ctx => ctx.sendToSrc(ctx.attr),
        (a, b) => a + b,
        TripletFields.EdgeOnly
    ).collect.toMap
    // normalize edges
    val normEdges = graph.edges.map(e => {
        val srcSum = outgoingEdgesSum.getOrElse(e.srcId,
            1.0)
        val newWeight = e.attr / srcSum
        Edge(e.srcId, e.dstId, newWeight)
    })
    val initialGuess = 1.0
    val vertices = graph.vertices
    // assign each vertex an initial guess
    Graph(vertices, normEdges).mapVertices((_, attr) =>
        initialGuess)
}
```

Listing 5.23: Preparing the graph for PageRank computation.

As described earlier, TextRank leverages the PageRank algorithm to extract a summary built as the juxtaposition of the top-k most significant sentences in the graph. As it takes into account the strength of the connection between a pair of sentences, proportional to their similarity, a modified version of the PageRank algorithm is needed, which can work with the weighted graph of sentences. As shown in Listing 5.24, the PageRank algorithm is implemented by using the Pregel APIs provided by GraphX, as discussed in Section 4.5.1.4. The only difference here lies in how the graph is prepared for the computation: In the standard case, the weight of each edge is normalized with respect to the out-degree of the source node, while in this case, it is normalized based on the sum of the weights of all the out-edges of the source node. Moreover, here the algorithm is not affected by the presence of sink nodes, as it computes PageRank on a fully connected graph of sentences. Once the PageRank is computed by running the Pregel operator for a fixed number of iterations, the *rankGraph* is obtained, which stores the corresponding PageRank value for each node. This graph is used within the *buildSummary* method (see Listing 5.25), together with the original graph of sentences, in order to extract the summary, which is finally printed on the console output.

```scala
def main(args: Array[String]) = {
    // build sentence graph
    val path = "src/main/scala/graphX_apps/ANN.txt"
    val sentenceGraph = buildGraph(path)
    // prepare graph for PageRank computation
    val preparedGraph = prepareGraph(sentenceGraph)
    // Define Pregel's UDFs for PageRank
    val d = 0.85
    val numVertices = preparedGraph.numVertices
    def vertexProgram(id: VertexId, attr: Double, msgSum:
        Double): Double = (1 - d) / numVertices + d * msgSum
    def sendMessage(edge: EdgeTriplet[Double, Double]):
        Iterator[(VertexId, Double)] = Iterator((edge.dstId,
        edge.srcAttr * edge.attr))
    def messageCombiner(a: Double, b: Double): Double = a +
        b
    // Execute Pregel for a fixed number of iterations.
```

```
    val rankGraph = Pregel(graph = preparedGraph, initialMsg
        = 0.0, maxIterations = 50)(vprog = vertexProgram,
        sendMsg = sendMessage, mergeMsg = messageCombiner)
    // Extract summary
    val top_k = 3
    val summary = buildSummary(sentenceGraph, rankGraph,
        top_k)
    println("Summary:\n" + summary)
}
```

Listing 5.24: Main method: PageRank UDFs definition and Pregel run.

Listing 5.25 shows how the summary is built starting from *rank-Graph*, the original graph of sentences, and an integer *k*, which indicates the number of sentences to be inserted into the summary. The summary will thus be composed of the top-*k* sentences by PageRank, ordered according to how they appear in the original text, where this ordering is kept by leveraging the node ID in the sentence graph.

```
def buildSummary(sentenceGraph: Graph[String, Double],
    rankGraph: Graph[Double, Double], k: Int): String ={
    // select the top-k sentences by PageRank, ordered
        according to how they appear in the original text
    sentenceGraph.vertices.join(rankGraph.vertices)
        .map { case (id, (sent, rank)) => (id, sent, rank) }
        .top(k)(Ordering.by(_._3)).sortBy(_._1)
        .map { case (_, sent, _) => sent }
        .mkString(".\n") + "."
}
```

Listing 5.25: Summary extraction starting with the rank graph and the original graph of sentences.

In the following, we report the output generated by the algorithm, starting from a text file containing a brief excerpt from Wikipedia describing artificial neural networks (ANNs).

Input (ANN.txt)

Artificial neural networks (ANNs), usually simply called neural networks (NNs) or neural nets, are computing systems inspired by the biological neural networks that constitute animal brains. An ANN is based on a collection of connected units or nodes called artificial neurons, which loosely model the neurons in a biological brain. Each connection, like the synapses in a biological brain, can transmit a signal to other neurons. An artificial neuron receives signals then processes them and can signal neurons connected to it. The signal at a connection is a real number, and the output of each neuron is computed by some nonlinear function of the sum of its inputs. The connections are called edges. Neurons and edges typically have a weight that adjusts as learning proceeds. The weight increases or decreases the strength of the signal at a connection. Neurons may have a threshold such that a signal is sent only if the aggregate signal crosses that threshold. Typically, neurons are aggregated into layers. Signals travel from the first layer (the input layer), to the last layer (the output layer), possibly after traversing the layers multiple times. The training of a neural network from a given example is usually conducted by determining the difference between the processed output of the network (often a prediction) and a target output. This difference is the error. The network then adjusts its weighted associations according to a learning rule and using this error value. Successive adjustments will cause the neural network to produce output that is increasingly similar to the target output. Such systems "learn" to perform tasks by considering examples, generally without being programmed with task-specific rules.

Output (summary)

An ANN is based on a collection of connected units or nodes called artificial neurons, which loosely model the neurons in a biological brain. The signal at a connection is a real number, and the output of each neuron is computed by some nonlinear function of the sum of its inputs. The training of a neural network from a given example is usually conducted by determining the difference between the processed output of the network (often a prediction) and a target output.

5.3.4.2 *MPI implementation*

In this section, we discuss the MPI implementation of the TextRank algorithm described above. We used the Java binding provided by Open MPI, an open-source implementation of MPI.

First of all, similarly to the previously described implementation, we defined a method to compute the textual similarity for a given pair of sentences (see Listing 5.26).

```java
public static double sentenceSimilarity(String sent1, String
    sent2) {
    Set<String> words1 = new HashSet<>(Arrays.asList(sent1.
        toLowerCase().split("[\\s,.;:?!]+")));
    Set<String> words2 = new HashSet<>(Arrays.asList(sent2.
        toLowerCase().split("[\\s,.;:?!]+")));
    int commonWords = (int) words1.stream().filter(words2::
        contains).count();
    double den = Math.log(words1.size() + 1) + Math.log(
        words2.size() + 1) + 1;
    return (double) commonWords / den;
}
```

Listing 5.26: Sentence similarity.

Listing 5.27 reads the input file and returns the list of all sentences it contains.

```
private static List<String> readInputFile(String fileName) {
    List<String> allSentences = new ArrayList<>();
    try (BufferedReader br = new BufferedReader(new
        FileReader(fileName))) {
        String line;
        while ((line = br.readLine()) != null) {
            String[] sentences = line.split("\\.");
            for (String s : sentences) {
                allSentences.add(s.strip());
            }
        }
    } catch (IOException e) {
        System.err.format("IOException: %s%n", e);
    }
    return allSentences;
}
```

Listing 5.27: Extracting all sentences from an input text file.

Starting from the list containing all sentences, the code in Listing 5.28 is used to build the fully connected graph of sentences. In particular, it is represented as a map with the *vertexId* of each sentence as the key and the set of in- and out-neighbors as the value. The neighborhoods of each vertex in the map are represented by a nested map that contains two keys, *in_neigh* and *out_neigh*, whose corresponding values are the lists of in-neighbor and out-neighbor IDs, respectively.

```
private static Map<Integer, Map<String, ArrayList<Integer>>>
    buildGraph(List<String> allSentences) {
    // store the graph as a Map using vertex\_id as the key
    // values are subMaps storing in- and out-neighborhoods
    Map<Integer, Map<String, ArrayList<Integer>>> graph =
        new HashMap<>();
    for (int i = 0; i < allSentences.size(); i++) {
        for (int j = 0; j < allSentences.size(); j++) {
            if (i != j) {
                if (!graph.containsKey(i)) {
                    // init neighborhoods of vertex i
                    graph.put(i, new HashMap<String,
                        ArrayList<Integer>>());
                    graph.get(i).put("in_neigh", new
                        ArrayList<Integer>());
```

```
            graph.get(i).put("out_neigh", new
                ArrayList<Integer>());
        }
        if (!graph.containsKey(j)) {
            // init neighborhoods of vertex j
            graph.put(j, new HashMap<String,
                ArrayList<Integer>>());
            graph.get(j).put("in_neigh", new
                ArrayList<Integer>());
            graph.get(j).put("out_neigh", new
                ArrayList<Integer>());
        }
        // add neighbors
        graph.get(i).get("out_neigh").add(j);
        graph.get(j).get("in_neigh").add(i);
        }
    }
}
    return graph;
}
```

Listing 5.28: Building the sentence graph starting with an input text file.

Once the methods described above had been defined, we leveraged MPI to implement the PageRank algorithm, which is the core component of the TextRank summarization technique. In the following, the main parts of the developed implementation are reported and discussed, focusing on the use of MPI.

First, we initialize the MPI execution environment by invoking the MPI.Init(args) method. Then, the size (i.e., the number of processes) and rank (i.e., the ID) of the current process are obtained as follows:

```
int size = MPI.COMM_WORLD.getSize();
int rank = MPI.COMM_WORLD.getRank();
```

Afterward, the list of all sentences is computed from the input file, which is used to create the sentence graph (see Listing 5.29). The number of vertices N is determined, and each node is assigned an initial guess equal to $\frac{1}{N}$. The PageRank value of the vertex with index i will be stored in the ith position of the *pageranks* array.

```
List<String> allSentences = readInputFile("ANN.txt");
Map<Integer, Map<String, ArrayList<Integer>>> graph =
    buildGraph(allSentences);
int numVertices = graph.keySet().size();
double[] pageranks = new double[numVertices];
Arrays.fill(pageranks, 1.0 / numVertices);
```

Listing 5.29: Sentence graph creation from input file.

Then, we compute the number of vertices that each process will handle and the remainder to handle the cases in which the number of vertices is not a multiple of the number of processes. Starting from this, we divide the whole set of vertices into a number of partitions equal to the *size* variable, which specifies the number of processes, by assigning each partition to a different process. In particular, we compute the range of indexes indicating the vertices for which each process will compute the PageRank value, thus distributing the PageRank computation of the whole set of nodes across different ranks in a data-parallel fashion. It is worth noting that, in the case where the remainder is greater than 0, the remaining nodes are assigned to the last process (see Listing 5.30).

```
// Compute the number of vertices that each process will
    handle
int verticesPerProcess = numVertices / size;
int remainder = numVertices % size;
// Compute the starting and ending indices for this process
int startIndex = rank * verticesPerProcess;
int endIndex = startIndex + verticesPerProcess;
if (remainder > 0 && rank == size - 1) {
    endIndex += remainder;
    verticesPerProcess += remainder;
}
```

Listing 5.30: Divide the data by assigning each rank a set of vertices to work with. If present, the remainder is handled by the last rank.

Listings 5.31 and 5.32 show how, for a given rank, a PageRank iteration is computed. Specifically, as shown in Listing 5.31, if convergence is not yet reached, the process initializes an array in which the local PageRank value of the nodes assigned to it will be stored. Then, for each assigned vertex v, the process performs the following operations:

- For each in-neighbor i of v, it computes the similarity between v and i, i.e., the weight of the edge connecting the vth vertex and the ith sentence in the *allSentences* list.
- The weight of the edge connecting v and i is then normalized with respect to the sum of the weights of all the out-edges of i.
- The contribution of vertex i to the local PageRank value of vertex v is summed up into the *localPageRanks* array.
- The overall contribution related to node v, stored in the *localPageranks* array, is finally converted into the actual PageRank value by applying the formula discussed in Section 4.5.1.4.

```
// stores the global PageRank values for each vertex of the
   graph
double[] globalPageranks = new double[numVertices];
// iterate until convergence
boolean converged = false;
while (!converged) {
    double[] localPageranks = new double[verticesPerProcess
       ];
    double sum_out_sim = 0.0;
    double in_weight = 0.0;
    for (int i = startIndex; i < endIndex; i++) {
        for (int in_neighbor : graph.get(i).get("in_neigh"))
           {
            in_weight = sentenceSimilarity(allSentences.get(
               in_neighbor), allSentences.get(i));
            // normalize weights of incoming edges
            sum_out_sim = 0.0;
            for (int out_neighbor : graph.get(in_neighbor).
               get("out_neigh")) {
                sum_out_sim += sentenceSimilarity(
                   allSentences.get(in_neighbor),
                   allSentences.get(out_neighbor));
            }
            in_weight /= sum_out_sim;
            localPageranks[i - startIndex] += pageranks[
               in_neighbor] * in_weight;
        }
        // compute local PageRank for the current assigned
           node
```

```
         localPageranks[i - startIndex] = DAMPING_FACTOR *
            localPageranks[i - startIndex] + (1 -
            DAMPING_FACTOR) / numVertices;
   }
   ...
```

Listing 5.31: PageRank iteration for computing local values.

In Listing 5.32, global PageRanks are collected for each process by gathering the local PageRank values computed by all other processes. This operation is usually performed through the *allGather* method, which provides a many-to-many communication pattern. In particular, given a set of elements distributed across all processes, *allGather* will gather all of the elements for all the processes. This behavior can be seen as the execution of a *gather* per process, followed by a broadcast operation. However, in this case, the possible presence of a remainder does not allow the use of the gather operation, which imposes that each rank communicate the same amount of data. To address this issue, the *allGatherv* method can be used, which is a generalized version of *allGather* in which each rank can send a variable amount of data. To perform this operation, the following parameters are specified:

- *localPageranks*: It is the local array of the current rank, containing the values to be sent.
- *verticesPerProcess*: It contains the number of vertices assigned to the current rank. In this implementation, all the ranks handle the same number of vertices, except for the last one (i.e., the process whose rank is equal to `size-1`), which will also handle the remainder.
- *MPI.DOUBLE*: It refers to the datatype of each item in the send buffer.
- *globalPageranks*: It is the global array in which received values will be collected and stored.
- *rcvCount*: It is an array containing the amount of data (i.e., number of PageRank values) sent by the ith rank and received by the current rank.

- *displs*: It is an array that specifies the offset position in the global array, starting from which the data received by the *i*th rank will be stored.
- *MPI.DOUBLE*: It refers to the datatype of each item in the receive buffer.

Once the global PageRank values are collected, each rank computes the current convergence value as the absolute difference (i.e., the delta) between the currently computed PageRank value and that from the previous iteration, and communicates this delta to all the other ranks through an *allGather*. It is worth noting that, in this case, the *allGather* method can be used, as each rank sends to the others exactly one value (i.e., the delta), regardless of the number of vertices it handles. Once the current rank has received the delta values from all other ranks, it can check for convergence. In particular, convergence is reached when all deltas are less than a given convergence threshold. Finally, the PageRank values are updated with the global result of the current iteration, and the subsequent step is performed unless convergence was reached.

```
  . . .

// Gather the local PageRank values from all processes
     handling varying numbers of vertices per rank
globalPageranks = new double[numVertices];
int[] rcvCount = new int[size];
int[] displs = new int[size];
for (int i = 0; i < size; i++) {
    rcvCount[i] = numVertices / size;
    displs[i] = rcvCount[i] * i;
}
rcvCount[size - 1] += remainder;
MPI.COMM_WORLD.allGatherv(localPageranks,
    verticesPerProcess, MPI.DOUBLE, globalPageranks,
    rcvCount, displs, MPI.DOUBLE);
// Compute the global PageRank variation for the
     iteration
double delta = 0.0;
for (int i = startIndex; i < endIndex; i++) {
    delta += Math.abs(globalPageranks[i] -
        pageranks[i]);
}
double[] deltas = new double[size];
```

```
MPI.COMM_WORLD.allGather(new double[] { delta }, 1, MPI.
    DOUBLE, deltas, 1, MPI.DOUBLE);
// Check convergence
converged = true;
for (double d : deltas) {
    if (d > CONVERGENCE_THRESHOLD) {
        converged = false;
        break;
    }
}
// Update the PageRank values for the next iteration
pageranks = globalPageranks;
}
```

Listing 5.32: Gathering of all local PageRank values and convergence check.

It is interesting to note that the convergence calculation could have been done using a single rank, replacing the *allGather* method with the *gather* primitive. However, reducing the communication from many-to-many to many-to-one would necessitate communicating the result of the convergence calculation (i.e., the convergence flag) back to all ranks. Furthermore, before performing the next iteration, each rank would have to wait for the arrival of the convergence flag, which would result in a lowering of the level of parallelism due to synchronization.

Finally, in Listing 5.33, the process with rank 0 computes the summary from the top three sentences by PageRank through the *extractSummary* method (see Listing 5.34), prints it in the console output, and terminates the MPI execution environment by invoking the MPI.Finalize() method.

```
if (rank == 0) {
    // rank 0 is responsible for extracting and printing
        the summary
    int topK = 3;
    String summary = extractSummary(globalPageranks,
        topK, allSentences);
    System.out.println("SUMMARY:\n" + summary);
}
MPI.Finalize();
```

Listing 5.33: Process with rank 0 computes the summary from the top three sentences by PageRank.

Listing 5.34 shows how the summary is extracted, starting from the array containing the final PageRank values for each sentence, the list of all sentences contained in the input text, and an integer k indicating the number of sentences to be included in the extracted summary. The output produced by this version is exactly the same as that generated by the GraphX-based implementation (see Section 5.3.4).

```java
public static String extractSummary(double[]
    globalPageranks,
  int k, List<String> allSentences) {
    Map<Double, Integer> map = new HashMap<>();
    for (int i = 0; i < globalPageranks.length; i++) {
        map.put(globalPageranks[i], i);
    }
    // get the top-k sentences by PageRank
    List<Double> values = new ArrayList<>(map.keySet());
    Collections.sort(values, Collections.reverseOrder());
    List<Integer> topKIndexes = new ArrayList<>();
    for (int i = 0; i < k && i < values.size(); i++) {
        topKIndexes.add(map.get(values.get(i)));
    }
    // order top-k sentences according to how they appear in
        the original text
    Collections.sort(topKIndexes);
    String summary = topKIndexes.stream().map(allSentences::
        get).collect(Collectors.joining(".\n")) + ".";
    return summary;
}
```

Listing 5.34: Summary extraction starting from the pagerank array and the list of sentences.

Comparing the two versions of the TextRank algorithm we provided, it is easy to see the higher verbosity of the MPI-based implementation with respect to the GraphX-based one discussed earlier (see Section 5.3.4.1). This is related to both the higher conciseness of the Scala language, derived from the use of functional programming, and the support for graph-based data structures and primitives. Indeed, the lack of such support in MPI necessitates the storage of different pieces of information related to vertices and neighborhoods using native data structures, which results in lower readability and higher verbosity of the code. This makes GraphX a more suitable choice for designing complex graph-parallel applications.

Chapter 6

Choosing the Right Framework to Tame Big Data

Throughout the chapters of this book, we introduced and analyzed the most important programming frameworks that have been developed for the implementation of scalable data-intensive distributed and/or parallel applications. In describing and comparing those frameworks, we focused on their features that allow developers to implement big data analysis and machine learning applications and analyzed how the discussed programming tools can be exploited to implement parallel operations in data-intensive algorithms and applications. This chapter concludes the book and discusses a set of factors that should be considered when developers select the appropriate framework for designing and implementing their big data applications. The lessons that can be learned from the analysis of the different programming tools we discussed in the book suggest that data features, application goals, and computing infrastructures should guide developers in selecting the best (or most appropriate) programming framework. This approach can be useful to guide designers and developers in appropriately selecting one or more of the discussed programming frameworks when they need to implement a given big data application or a specific machine learning technique that needs to access and process large datasets, providing scalability and high performance.

As mentioned, the main factors to be considered when implementing data-intensive applications include the following:

- *Input data characteristics*: It includes data volume, both in terms of size and dimensionality of the input dataset, data velocity, and data variety, which are important elements to be considered.
- *Application class*: It refers to the type of application that must be implemented and its goals; it could be a batch, a data streaming, a query-based, a graph-based, or another type of data processing application.
- *Hardware–software infrastructure*: It consists of the storage and computing infrastructure that will be used to run the designed application; for example an HPC platform, a cloud-based infrastructure (either public or private), or a server-based cluster, while considering issues related to data location and accessibility.

Along with these main factors, we should also consider other factors that generally influence the choices of developers and the application life cycle. Among them are:

- the programming skills of designers and developers;
- the features of the ecosystem of the selected programming framework;
- the size and activity degree of the community of developers;
- the data privacy requirements defining data access and processing;
- the costs of the hardware/software infrastructure to be used;
- the availability of data analysis and/or machine learning libraries that can be used in the programming framework;
- the abstraction level of the programming model offered by the framework.

In the following sections of the chapter, the above factors are discussed and analyzed, beginning with the three main ones.

6.1   Input Data

When addressing the challenges posed by big data, the selection of an appropriate programming framework is key to efficiently handling the volume, velocity, variety, and other characteristics of the data to be analyzed.

Volume, which refers to the amount of data generated or collected by applications, significantly affects the choice of the programming model and system from different perspectives. First, the volume impacts the storage requirements of the application, as storing large amounts of data requires distributed storage solutions that provide data replication, fault tolerance, and scalability. For this purpose, distributed file systems, such as Hadoop Distributed File System (HDFS), are designed to store and manage data across a cluster of machines, providing fault tolerance and scalability. Second, data volume also affects the processing requirements of an application, as processing large volumes of data requires distributed computing systems that can scale horizontally across a cluster of machines, such as Hadoop, which is commonly used for distributed and parallel processing of big data. On the other hand, Apache Spark offers in-memory processing capabilities, which make it suitable for iterative algorithms and interactive data analysis. Another challenge of processing high-volume data is related to the dimensions and features of the data. In many cases, high-dimensional data may require the use of dimensionality reduction techniques, such as principal component analysis (PCA) or singular value decomposition (SVD). However, only a few big data frameworks support these techniques, such as Spark through the MLlib library, which can make it easier to analyze and derive insights from high-dimensional data. Moreover, data volume also affects application performance. In fact, with the increase in data size, the time required to process the data increases, making it difficult to meet the performance requirements of the application. In such cases, choosing a programming system that supports both data and task parallelism, like Spark, could be appropriate to ensure the efficient processing of big data.

Velocity, which refers to the speed at which data are generated, is another important characteristic of big data that demands programming models and systems that can capture, process, and analyze data in (near) real time. Indeed, the velocity of the data impacts the data acquisition and processing requirements of applications. In high-speed data scenarios, data must be acquired and processed in real time to enable timely decision-making. While micro-batch streaming systems such as Spark Streaming can be used in some scenarios, stream processing frameworks, such as Apache Storm, are typically used to process these data streams in real time, allowing applications to respond to events as they occur. The speed of the data also affects the processing and analysis capabilities of the application since techniques such as windowing and time-based aggregation are required to capture relevant information from a data stream generated at high velocity. Furthermore, processing data in real time necessitates low-latency processing capabilities, such as in-memory processing, to enable the responsiveness of a big data application.

Variety, which refers to the heterogeneity of data types, formats, and sources generated or collected by an application, requires handling data integration, transformation, and analysis in a flexible and adaptable manner. Data from multiple sources and formats can be preprocessed before analysis using frameworks that support extract, transform, and load (ETL) operations, such as Apache Hive and Apache Pig. In addition, different data types require different analysis techniques. For example, structured data can be analyzed using traditional relational database techniques, such as those provided by Hive and Pig, while unstructured data (e.g., text) require special-purpose techniques (e.g., natural language processing (NLP) for textual data), which are provided by a few frameworks, including Apache Spark. Therefore, choosing a programming system that supports multiple data types and analysis techniques is critical to managing data diversity in big data applications. Among the frameworks discussed in this book, Spark is the most versatile for processing heterogeneous data since it provides APIs for batch processing, stream processing, machine learning, graph processing, and DataFrames to work with different data types, such as CSV files, JSON data, and database tables.

6.2   Application Class

Choosing the right programming solution for developing a big data analysis application significantly depends on the characteristics and requirements of the application a developer must implement. Some programming solutions are more general and can therefore be used profitably for different classes of applications, while others can be efficiently used in a specific field (e.g., stream data processing, data querying, and deep learning). In fact, general-purpose systems, such as Hadoop and Spark, are widely used due to their flexibility and ability to handle large datasets, whereas more specialized systems may offer advantages in performance and ease of use. Big data applications are used in many contexts and serve different purposes. They can be categorized into four main classes: batch, streaming, graph-based, and query-based.

Batch applications are designed to process and analyze large volumes of data in batch mode. This approach involves processing a large amount of data that is collected and analyzed together, typically during off-peak hours when the processing demand on the system is low. Examples of batch big data applications include data warehousing, data mining, and batch processing of large datasets. Batch data-intensive applications are particularly useful for analyzing historical data, generating reports, and performing complex analytics that require significant processing power and resources. Apache Spark and Apache Hadoop are both widely used for batch processing due to their distributed storage and processing capabilities. Hadoop includes fault-tolerant storage and a framework for distributed processing, while Spark offers a fast and flexible data processing engine with high-level APIs, machine learning, and more. Additionally, Apache Airflow can be used for developing and monitoring batch-oriented, workflow-based applications. In fact, it can be used to orchestrate and automate batch processing workflows, allowing final users to focus on generating meaningful insights from their data rather than managing the processing infrastructure.

Stream applications are designed to process and analyze data collected in real time. This approach involves processing data as they are received, without the need to store them in centralized repositories. Stream big data applications are particularly useful in application domains where real-time data analysis is required, such as finance, telecommunications, and transportation. Examples of stream big data applications include real-time data analytics, fraud detection, and real-time monitoring of sensors and IoT devices. Apache Storm and Apache Spark are both open-source frameworks designed for stream processing of data. Storm offers low-latency, scalable, and fault-tolerant real-time processing, while Spark provides micro-batch stream processing through specialized APIs. Both frameworks are suitable for processing large volumes of data in real time, making them ideal for stream processing applications that require fast and efficient analysis of data.

Graph-based applications are designed to process and analyze data that are interconnected in complex networks or graph structures. This approach involves analyzing relationships between different data nodes in a graph to uncover patterns and insights that may not be apparent when using traditional analysis methods. Examples of graph-based big data applications include social network analysis, recommendation engines, and fraud detection. Graph-based data-intensive applications are particularly useful in those cases where the data are highly interconnected and relationships between data points are critical to the analysis. MPI and Apache Spark are frameworks suitable for graph processing of big data. MPI offers low-level control over parallelism and communication, while Spark provides specialized high-level APIs (i.e., GraphX) for efficient and scalable graph processing. Both are suitable for analyzing complex relationships between nodes and processing large-scale graphs in a distributed and efficient manner.

Query-based applications are designed to provide fast and efficient access to large volumes of data through query languages and search tools. This approach involves storing data in a distributed system and using query languages, such as SQL, to retrieve data from the system. Query-based big data applications are particularly useful in applications where fast, *ad hoc* queries are required to extract insights from large datasets. Examples of query-based big data applications include business intelligence, data exploration, and *ad hoc* data analysis. Apache Hive, Pig, and Spark are open-source frameworks suitable for query processing of large datasets. Hive provides an SQL-like interface, Pig a simple scripting language, and Spark SQL allows querying and analyzing data using an SQL syntax. These frameworks are designed for fast and efficient access to large datasets and are ideal for *ad hoc* querying and analysis, such as data exploration and business intelligence.

6.3   Infrastructure

The emergence of big data technologies and applications has led to the development of complex infrastructures that enable efficient processing, storage, and analysis of vast amounts of data. Choosing the right infrastructure for big data analysis is critical for businesses and organizations to effectively harness their data and gain insights to drive decision-making. However, selecting the right framework that aligns with infrastructure requirements can be quite challenging since it depends on several factors, including infrastructure scale, big data use cases, data governance and security, cost and licensing, and scalability requirements. In this section, we discuss how, following these different factors, the most suitable big data framework can be selected according to the specific type of infrastructure used. In particular, we discuss recommendations for the main types of infrastructure choices, i.e., on-premise, cloud-based, and hybrid.

On-premise infrastructure refers to the deployment of hardware and software within an organization's premises, which does not require transferring large amounts of data outside to a remote location. This infrastructure is ideal for organizations that have strict data security requirements and want complete control over their data. Installing, configuring, and maintaining this type of infrastructure require a high level of expertise and a significant economic investment. When dealing with an on-premise infrastructure, data are processed and stored in a proprietary data center, which allows for higher security and easier compliance with stringent data accessibility and privacy regulations. On-premise infrastructures, especially for small- and medium-sized organizations with a limited IT budget, are often made up of interconnected machines equipped with commodity hardware. In these cases, Apache Hadoop can be effectively used to process large datasets at a lower cost on heterogeneous commodity hardware, relying on any disk storage type for data processing. Moreover, HDFS is capable of distributing data on different machines running different operating systems without requiring special drivers. For IT companies with larger budgets, Apache Spark is an effective solution for quick in-memory processing of large amounts of data. However, it operates at a higher cost because it requires large amounts of RAM to spin up nodes.

Cloud infrastructure refers to the use of cloud-based resources to store and process data. Cloud-based solutions are scalable and cost-effective, require less maintenance, and offer different types of services, including infrastructure as a service (IaaS), platform as a service (PaaS), and software as a service (SaaS). Cloud computing services are usually adopted for their scalability and flexibility, allowing them to add and remove resources based on application needs. Such services are usually offered by providers such as Amazon Web Services (AWS), Microsoft Azure, and Google Cloud Platform (GCP), and include specific services for big data processing, such as Amazon EMR, Azure HDInsight, Google Cloud Dataproc, and fully managed big data frameworks, such as Hadoop, Spark, and Flink, which are optimized for the cloud, providing scalability, ease of use, and cost-effective solutions without the need to manage a private infrastructure. In this way, organizations can focus on their own core processes and operations, offloading the burden of maintaining an up-to-date infrastructure that needs to be constantly updated and kept operational. However, the use of cloud infrastructures poses many privacy and data management issues, including those relating to security, regulatory compliance, jurisdictional constraints, and data access control. More in detail, since data regulations, such as the General Data Protection Regulation (GDPR) and the California Consumer Privacy Act (CCPA), require that organizations implement measures to protect their data and those of their clients, it is critical that a public cloud infrastructure meet these regulations to avoid legal issues. Moreover, data regulations and privacy laws can vary significantly between countries and regions. Thus, when using a public cloud infrastructure, it is crucial to understand where your data are being stored and how they are being processed. It is also important to ensure that your data are isolated from other users and that the provider has robust access controls, so as to avoid data breaches and unauthorized access.

Hybrid infrastructure is a combination of computing, storage, and services in different environments, including on-premises and cloud-based ones. This infrastructure enables workload management between various environments, allowing the development of more versatile and flexible solutions. It also provides both privacy preservation — by locally storing and processing sensitive data — and the ability to adapt to the organization's needs through cloud-based pay-as-you-go services that allow us to scale, customize, and provision effectively. In hybrid infrastructures, where a combination of on-premise and cloud-based components is used, several frameworks, supporting hybrid data processing, can be a suitable choice. Apache Spark supports reading and writing data from several sources, including HDFS, Network File Systems (NFS), and many cloud object storage services (e.g., Amazon S3, Azure Blob Storage, and Google Cloud Storage). In such a way, it is possible to configure complex data pipelines in which data can be read from on-premise sources and then loaded, processed, and saved on the cloud. In many other cases, it is possible to exploit specific frameworks for data integration. For example, Apache Kafka is a distributed streaming platform that allows for building real-time data pipelines between on-premise and cloud-based environments and is often used as a data integration layer between these environments. Similarly, Apache Airflow can be used for data ingestion and processing, enabling data to flow seamlessly between on-premises and cloud-based environments.

6.4   Other Factors

Following the above-discussed main factors that need to be considered when selecting a big data programming framework for developing data-intensive applications, here we recall the other factors listed previously that generally influence, to some extent, the decisions of developers about the right programming framework to use. The additional factors that are worthwhile to consider are as follows:

- *Skills of designers and developers*: Programmers generally start with their coding expertise; however, although high-level programming tools are preferred, a good knowledge of low-level programming mechanisms may help in achieving good performance for the developed applications.
- *Programming framework ecosystem*: A rich set of tools linked to the selected framework is welcome; in fact, if the programming framework is part of a tool ecosystem the developer's productivity may increase.
- *Size of the community*: The more developers, maintainers, and users of the programming environment you want to use, the easier it is to obtain support in the development life cycle.
- *Data privacy requirements*: As we mentioned in Section 6.1 on input data, requirements about data privacy are important because they govern how data can be accessed, analyzed, and exchanged in the application to be developed.

- *Costs of the hardware/software infrastructure*: The selection of the appropriate programming framework must sometimes also consider the costs of the hardware infrastructure where the application will run after its implementation.
- *Availability of additional libraries*: The ease of access to data analysis and/or machine learning libraries that are integrated into the selected programming framework can be a helpful element to consider when complex analysis or learning algorithms must be implemented.
- *Abstraction level*: Although a good knowledge of low-level programming may help increase application performance, the selection of a high-level programming model allows developers to abstract from low-level parallelism or distribution details and makes it easier and faster to implement applications.

Supplementary Material

The supplementary material includes:

(a) lecture slides in PowerPoint;
(b) lecture slides in PDF.

Online access is automatically assigned if you purchase the ebook online via www.worldscientific.com.

If you have purchased the print copy of this book or the ebook via other sales channels, please follow the instructions below to download the files:

1. Go to: https://www.worldscientific.com/r/q0444-supp or scan the QR code below.

258 Programming Big Data Applications: Scalable Tools and Frameworks

2. Register an account/login.
3. Download the files from: https://www.worldscientific.com/world scibooks/10.1142/q0444#t=suppl.

For subsequent access, simply log in with the same login details in order to access.

For enquiries, please email: sales@wspc.com.sg.

Bibliography

Abramova, V., Bernardino, J., and Furtado, P. (2014). Which NoSQL database? A performance overview, *Open Journal of Databases (OJDB)* **1**(2), 17–24.

Agrawal, D., Bernstein, P., Bertino, E., Davidson, S., Dayal, U., Franklin, M., Gehrke, J., Haas, L., Halevy, A., Han, J., *et al.* (2012). Challenges and Opportunities with Big Data. A community white paper developed by leading researchers across the United States (Computing Research Association, Washington).

Agrawal, R., and Shafer, J. (1996). Parallel mining of association rules, *IEEE Transactions on Knowledge and Data Engineering* **8**(6), 962–969.

Barga, R., Gannon, D., and Reed, D. (2011). The client and the cloud: Democratizing research computing, *IEEE Internet Computing* **15**(1), 72–75.

Bauer, M., Treichler, S., Slaughter, E., and Aiken, A. (2012). Legion: Expressing locality and independence with logical regions, in *Proceedings of the International Conference on High Performance Computing, Networking, Storage and Analysis*, SC '12 (IEEE Computer Society Press, Washington, DC, USA). ISBN 9781467308045.

Bekkerman, R., Bilenko, M., and Langford, J. (2011). *Scaling up Machine Learning: Parallel and Distributed Approaches* (Cambridge University Press).

Belcastro, L., Marozzo, F., Talia, D., and Trunfio, P. (2017). Big data analysis on clouds, in A. Zomaya and S. Sakr (eds.), *Handbook of Big Data Technologies* (Springer), pp. 101–142. ISBN: 978-3-319-49339-8.

Belcastro, L., Marozzo, F., Talia, D., and Trunfio, P. (2018). G-RoI: Automatic region-of-interest detection driven by geotagged social media data, *ACM Transactions on Knowledge Discovery from Data* **12**(3), 27:1–27:22.

Belcastro, L., Marozzo, F., and Talia, D. (2019). Programming models and systems for big data analysis, *International Journal of Parallel, Emergent and Distributed Systems* **34**, 632–652.

Belcastro, L., Cantini, R., Marozzo, F., Talia, D., and Trunfio, P. (2020). Learning political polarization on social media using neural networks, *IEEE Access* **8**(1), 47177–47187.

Belcastro, L., Marozzo, F., and Perrella, E. (2021a). Automatic detection of user trajectories from social media posts, *Expert Systems with Applications* **186**, 115733.

Belcastro, L., Marozzo, F., Talia, D., and Trunfio, P. (2021b). Cloud computing for enabling big data analysis, in *The 10th International Conference on Cloud Computing and Services Science (CLOSER 2020)*, pp. 84–109. ISBN 978-3-030-72369-9.

Belcastro, L., Cantini, R., Marozzo, F., Orsino, A., Talia, D., and Trunfio, P. (2022). Programming big data analysis: Principles and solutions, *Journal of Big Data* **9**(4), 1–50.

Bell, G., Hey, T., and Szalay, A. (2009). Beyond the data deluge, *Science* **323**(5919), 1297–1298.

Bergman, K., Borkar, S., Campbell, D., Carlson, W., Dally, W., Denneau, M., Franzon, P., Harrod, W., Hill, K., Hiller, J., *et al.* (2008). Exascale computing study: Technology challenges in achieving exascale systems, Defense Advanced Research Projects Agency Information Processing Techniques Office (DARPA IPTO), Technical Report, Vol. 15, p. 181.

Beyer, M. A. and Laney, D. (2012). The importance of 'big data': A definition (Gartner, Stamford, CT), pp. 2014–2018.

Brin, S., and Page, L. (1998). The anatomy of a large-scale hypertextual web search engine, *Computer Networks and ISDN Systems* **30**(1–7), 107–117.

Brown, S. D., Francis, R. J., Rose, J., and Vranesic, Z. G. (1992). *Field-Programmable Gate Arrays*, Vol. 180 (Springer Science & Business Media).

Buyya, R. (1999). *High Performance Cluster Computing*, Vol. 2 (New Jersey: Prentice).

Byun, C., Arcand, W., Bestor, D., Bergeron, B., Gadepally, V., Houle, M., Hubbell, M., Jananthan, H., Jones, M., Keville, K., Klein, A., Michaleas, P., Milechin, L., Morales, G., Mullen, J., Prout, A., Reuther, A., Rosa, A., Samsi, S., Yee, C., and Kepner, J. (2022).

pPython for parallel python programming, in *2022 IEEE High Performance Extreme Computing Conference (HPEC)*, pp. 1–6.

Cantini, R., Marozzo, F., Orsino, A., Talia, D., and Trunfio, P. (2021). Exploiting machine learning for improving in-memory execution of data-intensive workflows on parallel machines, *Future Internet* **13**(5), 121.

Cao, L. (2017). Data science: A comprehensive overview, *ACM; Computing Surveys* **50**(3–43), 1–42.

Cattell, R. (2011). Scalable SQL and NoSQL data stores, *ACM SIGMOD Record* **39**(4), 12–27.

Cesario, E., Marozzo, F., Talia, D., and Trunfio, P. (2017). SMA4TD: A social media analysis methodology for trajectory discovery in large-scale events, *Online Social Networks and Media* **3–4**, 49–62.

Chang, F., Dean, J., Ghemawat, S., Hsieh, W. C., Wallach, D. A., Burrows, M., Chandra, T., Fikes, A., and Gruber, R. E. (2008). Bigtable: A distributed storage system for structured data, *ACM Transactions on Computer Systems (TOCS)* **26**(2), 4.

Chang, W. and Grady, N. (2015). NIST big data interoperability framework: Volume 1, big data definitions, Special Publication (NIST SP), National Institute of Standards and Technology, Gaithersburg, MD.

Charles, P., Grothoff, C., Saraswat, V., Donawa, C., Kielstra, A., Ebcioglu, K., Von Praun, C., and Sarkar, V. (2005). X10: An object-oriented approach to non-uniform cluster computing, *ACM SIGPLAN Notices* **40**(10), 519–538.

Da Costa, G., Fahringer, T., Rico-Gallego, J.-A., Grasso, I., Hristov, A., Karatza, H., Lastovetsky, A., Marozzo, F., Petcu, D., Stavrinides, G., Talia, D., Trunfio, P., and Astsatryan, H. (2015). Exascale machines require new programming paradigms and runtimes, *Supercomputing Frontiers and Innovations: International Journal* **2**(2), 6–27.

De Mauro, A., Greco, M., and Grimaldi, M. (2015). What is big data? A consensual definition and a review of key research topics, in *AIP Conference Proceedings*, Vol. 1644 (American Institute of Physics), pp. 97–104.

De Wael, M., Marr, S., De Fraine, B., Van Cutsem, T., and De Meuter, W. (2015). Partitioned global address space languages, *ACM Computing Surveys* **47**(4), 1–27.

Dean, J. and Ghemawat, S. (2004). MapReduce: Simplified data processing on large clusters, in *Proceedings of the 6th Conference on Symposium on Operating Systems Design & Implementation*, OSDI'04, Vol. 6 (USENIX Association, USA), p. 10.

Deelman, E., Singh, G., Su, M.-H., Blythe, J., Gil, Y., Kesselman, C., Mehta, G., Vahi, K., Berriman, G. B., Good, J., *et al.* (2005). Pegasus:

A framework for mapping complex scientific workflows onto distributed systems, *Scientific Programming* **13**(3), 219–237.

Deitz, S. J., Chamberlain, B. L., and Hribar, M. B. (2006). Chapel: Cascade high-productivity language an overview of the chapel parallel programming model, Cray User Group.

Dijcks, J.-P. (2013). *Oracle: Big Data for the Enterprise* (Oracle Corporation, USA).

Dongarra, J. J., Moler, C. B., Bunch, J. R., and Stewart, G. W. (1979). *LINPACK Users' Guide* (SIAM).

Flynn, M. J. (1972). Some computer organizations and their effectiveness, *IEEE Transactions on Computers* **100**(9), 948–960.

Flynn, M. J. and Rudd, K. W. (1996). Parallel architectures, *ACM Computing Surveys (CSUR)* **28**(1), 67–70.

Fuerlinger, K., Fuchs, T., and Kowalewski, R. (2016). DASH: A C++ PGAS library for distributed data structures and parallel algorithms, in *2016 IEEE 18th International Conference on High Performance Computing and Communications* (IEEE, Sydney, Australia), pp. 983–990.

Gagliardi, F., Moreto, M., Olivieri, M., and Valero, M. (2019). The international race towards exascale in Europe, *CCF Transactions on High Performance Computing* **1**(1), 3–13.

Gajendran, S. K. (2012). A Survey on NoSQL Databases (University of Illinois).

Ganesh Chandra, D. (2015). Base analysis of NoSQL database, *Future Generation Computer Systems* **52**, 13–21 (Special Section: Cloud Computing: Security, Privacy and Practice).

Gantz, J. and Reinsel, D. (2011). Extracting value from chaos, *IDC iView* **1142**, 1–12.

Gartner, Inc. (nd). What is big data? Gartner IT Glossary, http://www.gartner.com/it-glossary/big-data.

Gates, A. F., Natkovich, O., Chopra, S., Kamath, P., Narayanamurthy, S. M., Olston, C., Reed, B., Srinivasan, S., and Srivastava, U. (2009). Building a high-level dataflow system on top of Map-Reduce: The Pig experience, *Proceedings of the VLDB Endowment* **2**(2), 1414–1425.

Geist, A., Gropp, W., Huss-Lederman, S., Lumsdaine, A., Lusk, E., Saphir, W., Skjellum, T., and Snir, M. (1996). MPI-2: Extending the message-passing interface, in *Euro-Par'96 Parallel Processing* (Springer), pp. 128–135.

Gilbert, S. and Lynch, N. (2002). Brewer's conjecture and the feasibility of consistent, available, partition-tolerant web services, *ACM SIGACT News* **33**(2), 51–59.

Grama, A., Gupta, A., Karypis, G., and Kumar, V. (2003). *Introduction to Parallel Computing* (Addison Wesley, Harlow, England).

Gropp, W. and Snir, M. (2013). Programming for exascale computers, *Computing in Science & Engineering* **15**(6), 27–35.

Han, J., Pei, J., and Yin, Y. (2000). Mining frequent patterns without candidate generation, *SIGMOD Record* **29**(2), 1–12.

Hashem, I. A. T., Yaqoob, I., Anuar, N. B., Mokhtar, S., Gani, A., and Khan, S. U. (2015). The rise of "big data" on cloud computing: Review and open research issues, *Information Systems* **47**, 98–115.

Hey T., Tansley S., and Tolle, K. (2009). *The Fourth Paradigm* (Microsoft).

Huai, Y., Chauhan, A., Gates, A., Hagleitner, G., Hanson, E. N., O'Malley, O., Pandey, J., Yuan, Y., Lee, R., and Zhang, X. (2014). Major technical advancements in Apache Hive, in *Proceedings of the 2014 ACM SIGMOD International Conference on Management of Data*, pp. 1235–1246.

Kalé, L. and Krishnan, S. (1993). CHARM++: A portable concurrent object oriented system based on C++, in A. Paepcke (ed.), *Proceedings of OOPSLA'93* (ACM Press), pp. 91–108.

Kargupta, H., Park, B., Hershberger, D., and Johnson, E. (1999). Collective data mining: A new perspective toward distributed data mining, *Advances in Distributed and Parallel Knowledge Discovery* **2**, 131–174.

Kornacker, M., Behm, A., Bittorf, V., Bobrovytsky, T., Ching, C., Choi, A., Erickson, J., Grund, M., Hecht, D., Jacobs, M., *et al.* (2015). Impala: A modern, open-source SQL engine for Hadoop, in *CIDR*, Vol. 1, p. 9.

Kumar, A. and Sebastian, T. M. (2012). Sentiment analysis on Twitter, *International Journal of Computer Science Issues (IJCSI)* **9**(4), 372.

Laney, D. *et al.* (2001). 3d data management: Controlling data volume, velocity and variety, *META Group Research Note* **6**(70), 1.

Li, A., Yang, X., Kandula, S., and Zhang, M. (2010). CloudCmp: Comparing public cloud providers, in *Proceedings of the 10th ACM SIGCOMM Conference on Internet Measurement*, pp. 1–14.

Li, H., Wang, Y., Zhang, D., Zhang, M., and Chang, E. Y. (2008). PFP: Parallel FP-Growth for query recommendation, in *Proceedings of the 2008 ACM Conference on Recommender Systems*, RecSys '08 (Association for Computing Machinery, New York, NY, USA), pp. 107–114. ISBN 9781605580937.

Lordan, F., Tejedor, E., Ejarque, J., Rafanell, R., Alvarez, J., Marozzo, F., Lezzi, D., Sirvent, R., Talia, D., and Badia, R. M. (2014). ServiceSs: An interoperable programming framework for the cloud, *Journal of Grid Computing* **12**, 67–91.

Lourenço, J. R., Cabral, B., Carreiro, P., Vieira, M., and Bernardino, J. (2015). Choosing the right NoSQL database for the job: A quality attribute evaluation, *Journal of Big Data* **2**(1), 1–26.

Ludäscher, B., Altintas, I., Berkley, C., Higgins, D., Jaeger, E., Jones, M., Lee, E. A., Tao, J., and Zhao, Y. (2006). Scientific workflow management and the Kepler system, *Concurrency and Computation: Practice and Experience* **18**(10), 1039–1065.

Malewicz, G., Austern, M. H., Bik, A. J., Dehnert, J. C., Horn, I., Leiser, N., and Czajkowski, G. (2010). Pregel: A system for large-scale graph processing, in *Proceedings of the 2010 ACM SIGMOD International Conference on Management of Data* (ACM), pp. 135–146.

Marozzo, F., Talia, D., and Trunfio, P. (2012). P2P-MapReduce: Parallel data processing in dynamic cloud environments, *Journal of Computer and System Sciences* **78**(5), 1382–1402.

Marozzo, F., Talia, D., and Trunfio, P. (2018). A workflow management system for scalable data mining on clouds, *IEEE Transactions On Services Computing* **11**(3), 480–492. ISSN: 1939-1374.

Meijer, E. (2011). The World according to LINQ, *Communications of the ACM* **54**(10), 45–51.

Mell, P., Grance, T., *et al.* (2011). The NIST definition of cloud computing, Special Publication (NIST SP), National Institute of Standards and Technology, Gaithersburg, MD.

Mihalcea, R., and Tarau, P. (2004). TextRank: Bringing order into text, in *Proceedings of the 2004 Conference on Empirical Methods in Natural Language Processing* (Association for Computational Linguistics, Barcelona, Spain), pp. 404–411.

Moniruzzaman, A. B. M. and Hossain, S. A. (2013). NoSQL database: New era of databases for big data analytics — Classification, characteristics and comparison. *CoRR* **abs/1307.0191**.

Navarro, C. A., Hitschfeld-Kahler, N., and Mateu, L. (2014). A survey on parallel computing and its applications in data-parallel problems using GPU architectures, *Communications in Computational Physics* **15**(2), 285–329.

Nyce, C. and Cpcu, A. (2007). Predictive analytics white paper (American Institute for CPCU. Insurance Institute of America), pp. 9–10.

Olston, C., Reed, B., Silberstein, A., and Srivastava, U. (2008a). Automatic optimization of parallel dataflow programs.

Olston, C., Reed, B., Srivastava, U., Kumar, R., and Tomkins, A. (2008b). Pig Latin: A not-so-foreign language for data processing, in *Proceedings of the 2008 ACM SIGMOD international conference on Management of data*, pp. 1099–1110.

Otte, E. and Rousseau, R. (2002). Social network analysis: A powerful strategy, also for the information sciences, *Journal of Information Science* **28**(6), 441–453.

Owens, J. D., Houston, M., Luebke, D., Green, S., Stone, J. E., and Phillips, J. C. (2008). GPU computing, *Proceedings of the IEEE* **96**(5), 879–899.

Papamichail, M., Diamantopoulos, T., and Symeonidis, A. (2016). User-perceived source code quality estimation based on static analysis metrics, in *2016 IEEE International Conference on Software Quality, Reliability and Security (QRS)*, pp. 100–107.

Prodromidis, A., Chan, P., Stolfo, S., *et al.* (2000). Meta-learning in distributed data mining systems: Issues and approaches, *Advances in Distributed and Parallel Knowledge Discovery* **3**, 81–114.

Richardson, L. and Ruby, S. (2008). *RESTful Web Services* (O'Reilly Media, Inc.).

Rodriguez, M. A. and Neubauer, P. (2010). The graph traversal pattern, *CoRR* **abs/1004.1001**.

Salloum, S., Dautov, R., Chen, X., Peng, P. X., and Huang, J. Z. (2016). Big data analytics on Apache Spark, *International Journal of Data Science and Analytics* **1**(3), 145–164.

Sarkar, A., Ghosh, A., and Nath, D. A. (2015). MapReduce: A comprehensive study on applications, scope and challenges, *Department of Computer Science, International Journal of Advance Research in Computer Science and Management Studies* **3**(7).

Schroeck, M., Shockley, R., Smart, J., Romero-Morales, D., and Tufano, P. (2012). Analytics: The real-world use of big data, *IBM Global Business Services*, 1–20.

Singh, D. and Reddy, C. K. (2015). A survey on platforms for big data analytics, *Journal of Big Data* **2**(1), 1–20.

Skillicorn, D. B. and Talia, D. (1998). Models and languages for parallel computation, *ACM Computing Surveys (CSUR)* **30**(2), 123–169.

Spezzano, G. and Talia, D. (1999). *Calcolo parallelo, automi cellulari e modelli per sistemi complessi*, (FrancoAngeli).

Stonebraker, M. (2010). SQL databases v. NoSQL databases, *Communications of the ACM* **53**(4), 10–11.

Strohmaier, E., Meuer, H. W., Dongarra, J., and Simon, H. D. (2015). The TOP500 list and progress in high-performance computing, *Computer* **48**(11), 42–49.

Talia, D. (2013). Workflow systems for science: Concepts and tools, *International Scholarly Research Notices* **2013**.

Talia, D. (2019). A view of programming scalable data analysis: From clouds to exascale, *Journal of Cloud Computing* **8**(1), 1–16.

Talia, D. and Trunfio, P. (2012). *Service-Oriented Distributed Knowledge Discovery* (Chapman and Hall/CRC).

Talia, D., Trunfio, P., and Marozzo, F. (2015). *Data Analysis in the Cloud* (Elsevier). ISBN 978-0-12-802881-0.

Talia, D., Trunfio, P., Carrettero, J., and Garcia-Blas, J. (2022). Editorial research topic towards exascale solutions for big data computing, *Frontiers in Big Data*, 4.

Talia, D., Trunfio, P., Marozzo, F., Belcastro, L., Garcia-Blas, J., Rio, D. D., Couvée, P., Goret, G., Vincent, L., Fernández-Pena, A., *et al.* (2019). A novel data-centric programming model for large-scale parallel systems, in *European Conference on Parallel Processing* (Springer), pp. 452–463.

Tan, P.-N., Steinbach, M., and Kumar, V. (2016). *Introduction to Data Mining* (Pearson Education India).

Tiskin, A. (1998). The bulk-synchronous parallel random access machine, *Theoretical Computer Science* **196**(1–2), 109–130.

UPC Consortium (2005). UPC language specifications, v1. 2, Lawrence Berkeley National Lab tech report lbnl-59208, Technical Report, Berkeley, CA, USA.

Valiant, L. G. (1990). A bridging model for parallel computation, *Communications of the ACM* **33**(8), 103–111.

Van der Aalst, W. M. P., ter Hofstede, A. H. M., Kiepuszewski, B., and Barros, A. P. (2003). Workflow patterns, *Distributed and Parallel Databases* **14**(1), 5–51.

Van der Aalst, W. M., and ter Hofstede, A. H. (2005). Yawl: Yet another workflow language, *Information Systems* **30**(4), 245–275.

Verma, A., Mansuri, A. H., and Jain, N. (2016). Big data management processing with Hadoop MapReduce and Spark technology: A comparison, in *2016 Symposium on Colossal Data Analysis and Networking (CDAN)* (IEEE), pp. 1–4.

Vukotic, A., Watt, N., Abedrabbo, T., Fox, D., and Partner, J. (2015). *Neo4j in Action* (Manning).

Wadkar, S., Siddalingaiah, M., and Venner, J. (2014). *Pro Apache Hadoop* (Apress).

Wu, D., Sakr, S., and Zhu, L. (2017). Big data programming models, in *Handbook of Big Data Technologies* (Springer), pp. 31–63.

Zheng, Y., Kamil, A., Driscoll, M. B., Shan, H., and Yelick, K. (2014). UPC++: A PGAS extension for C++, in *2014 IEEE 28th International Parallel and Distributed Processing Symposium*, pp. 1105–1114.

Zikopoulos, P., Eaton, C., deRoos, D., Deutch, T. and Lapis, G. (2011). *Understanding Big Data: Analytics for Enterprise Class Hadoop and Streaming Data*, 1st edn. (McGraw-Hill Osborne Media). ISBN 0071790535, 9780071790536.

Index

www.ingramcontent.com/pod-product-compliance
Lightning Source LLC
Chambersburg PA
CBHW050545190326
41458CB00007B/1930